Gabriel G. Pessoa de Castro
Gustavo B. e Silva
Isabela T. Magacho

Solar-powered Stirling Engine

Gabriel G. Pessoa de Castro
Gustavo B. e Silva
Isabela T. Magacho

Solar-powered Stirling Engine

Development and Application

ScienciaScripts

Imprint

Any brand names and product names mentioned in this book are subject to trademark, brand or patent protection and are trademarks or registered trademarks of their respective holders. The use of brand names, product names, common names, trade names, product descriptions etc. even without a particular marking in this work is in no way to be construed to mean that such names may be regarded as unrestricted in respect of trademark and brand protection legislation and could thus be used by anyone.

Cover image: www.ingimage.com

This book is a translation from the original published under ISBN 978-613-9-78182-9.

Publisher:
Sciencia Scripts
is a trademark of
Dodo Books Indian Ocean Ltd. and OmniScriptum S.R.L publishing group

120 High Road, East Finchley, London, N2 9ED, United Kingdom
Str. Armeneasca 28/1, office 1, Chisinau MD-2012, Republic of Moldova, Europe
Printed at: see last page
ISBN: 978-620-6-65227-4

Authors

Msc. Gabriel Gonçalves Pessoa de Castro

University Professor at the Dom Bosco Educational Association - School of Engineering of
Resende has a bachelor's degree in mechanical engineering, a master's degree in metallurgical engineering
and is studying for a doctorate
in metallurgical engineering at the Fluminense Federal University (UFF), specialising in
occupational safety engineering at the Cruzeiro do Sul University.

Engineer Gustavo Borges e Silva

Mechanical Engineer graduated from the Dom Bosco Educational Association -
Resende School of Engineering.

Engineer Isabela Tereza Magacho

Mechanical Engineer graduated from the Dom Bosco Educational Association -
Resende School of Engineering.

DEDICATORY

To God who has enlightened me throughout this journey, to my family and friends who have been with me through good times and bad, being fundamental in this journey. I dedicate this monograph to you

Gustavo Borges E Silva

I dedicate this work to my family and friends who have always been by my side, giving me the strength to carry on. Without you I wouldn't have got this far. I am eternally grateful!

Isabela Tereza Magacho

ACKNOWLEDGEMENTS

Thanks first of all to God, without his faith I don't know what would become of me. He gave me life, a marvellous family, and I can't forget to mention my parents Tânia Borges and João Evangelista, my sister Juliana Borges and my grandmother Maria das Graças, who have been by my side for everything and have spared no effort to live this dream together with me, being with me in my joys, vibrations, fears, regrets, giving me strength whenever I needed it, both psychologically and financially, I will be eternally grateful. I would also like to thank my friends, and all the people who, in small gestures, helped me to achieve this feat, even though they had no knowledge of the subject, passing on positive energy, enlightening me to seek more knowledge and, above all, never giving up. This achievement is not just mine, but all of yours. Thank you.

Gustavo Borges E Silva

First of all, I would like to thank God, who gave me the opportunity to be here experiencing this achievement, my parents and grandparents who were by my side at all times giving me strength, my sister who gave up sleeping in our bedroom for many nights so that I could keep the lights on studying, friends who I was able to share such happy times with and who gave me the strength to carry on. TV Rio Sul, the company I work for, who believed in me and helped me financially, contributing greatly to my personal and professional growth. To the teachers and professors I've had the opportunity to meet on this journey and who have taught me so much. I would also like to express my gratitude to everyone who bought my sweets, you were fundamental to the realisation of this dream. To everyone involved, thank you very much. This victory is ours!

Isabela Tereza Magacho

SUMMARY

The constant evolution of machinery and equipment has caused great concern for the country's sustainable development. More and more industries, greater use of fossil fuels and, consequently, greater wear and tear on the environment. In view of the rapid advance of technology and the environmental impact it can have, this study aims to implement energy efficiency measures in a way that does not harm the environment, promoting a better quality of life. It is in this context that this project is an alternative way of producing electricity in an innovative way. The project consists of a Stirling engine that, coupled with a parabolic reflector antenna, is able to concentrate the sun's rays on a single point of the engine, the hot chamber. This makes it possible to generate a temperature difference between the two chambers and transform the solar energy captured into mechanical energy. Through the dynamo connected to the motor, this mechanical energy is transformed into electrical energy, in a clean and sustainable way. The project was designed using SolidWorks as a tool, which enabled the operation of the components to be analysed assertively. Through related research, it was possible to stipulate the performance according to the laws of thermodynamics and the economic and environmental aspects, presenting the feasibility of the project.

Keywords: sustainable development, environmental impact, Stirling engine, solar energy.

CHAPTER 1

INTRODUCTION

1.1   INITIAL CONSIDERATIONS

The environment is fundamental to the life of every human being, and sustainable development has therefore been given great prominence and importance. Society is currently experiencing a fourth industrial revolution, where manufacturing and information technology go hand in hand, and in the face of so much innovation, it is also necessary to define efficient methods of environmental preservation.

The function of the solar-powered Stirling engine is to generate energy efficiently, innovatively and using heat sources that are free of polluting gases.

The use of solar energy as a renewable source is very attractive and has numerous benefits, such as: it is an inexhaustible resource, highly powerful due to the large amount of radiation emitted daily and can be used to generate electricity.

1.2   PROBLEM SITUATION

As more and more super-powered machine designs are developed, the high demand for fossil fuels directly affects the environment, contributing to the greenhouse effect and influencing the quality of life. Climate change, catastrophic natural phenomena, global overheating, rising sea levels and the Earth's future resources are being used up in the present, leading to ecological deficits and causing great concern. Another problem encountered is the considerable increase in electricity tariffs, which are made up of various factors. Based on these factors, the feasibility of the Stirling engine for use in electricity generation was analysed, using solar heating as the heat source.

1.3   JUSTIFICATION FOR THE TOPIC

The search for ways to develop with a view to environmental safety is of great importance, as we live in a world rich in clean and efficient sources that can replace fossil fuels, reducing the impact they can have.

> The concept of sustainable development proposes a new economic and social order, at a planetary level, resulting from critical and reflective analyses of the historical relationship between human beings and the earth. The sustainability of development

is the most important concept to emerge from the debate on the environmental issue, because it has politically internalised ecology as a planning tool, opening up new perspectives for development and progress, as well as recovering human values and ethics that have been shattered by the absurd principles of traditional economics.

(NEGRET apud FERNANDES, 2004, p.57).

With the development of the solar-powered Stirling engine, it is possible to generate electricity in a sustainable and innovative way.

CHAPTER 2

OBJECTIVES

The aim of the work can be categorised into a general aim and a specific aim, which will be explained below in the monograph.

2.1   GENERAL OBJECTIVE

The theoretical development of a Stirling engine coupled to a low-scale parabolic trough collector for a power generation system using solar energy as the main source of heat.

2.2   SPECIFIC OBJECTIVES

- Studying the performance of a solar-powered Stirling engine;
- Study the thermodynamic aspects of the Stirling engine;
- Study the economic aspects of the Stirling engine;
- Study the principles of the parabolic concentrator.

CHAPTER 3

METHODOLOGY

In order to carry out this study, a bibliographical survey was carried out on subjects related to the Stirling engine. Theories, objectives, analyses and perspectives were explored, as well as technical manuals and scientific articles that dealt specifically with the topic presented.

The research took into account the history of the engine and the reason for its creation. Performance, applicability and its importance in reducing the environmental problems caused by the widespread use of fossil fuels.

During the study, weekly meetings were held to gather technical information on the engine, visits were made to machining companies, market research was carried out to analyse value, and project development was improved usingCatia and SolidWorks software.

CHAPTER 4

LITERATURE REVIEW

4.1 THE INDUSTRIAL REVOLUTION AND THE DEVELOPMENT OF TECHNOLOGY

The Industrial Revolution can be summed up as a period in which the transition from artisanal labour to the use of machines took place between the 18th and 19th centuries in Europe. Machines were developed with the aim of increasing production, consequently profits and minimising human effort.

"Steam power was adopted, but as the demands were not great, until 1838 a quarter of the energy was still from hydraulic sources" (Hobsbawm, 1968, p. 56).The steam engine was the main means used to move other machines, and its basic operation is based on three parts: the boiler where the steam is produced, the thermal machine that transforms the energy of the steam into mechanical energy, and the tender that transports the fuel and water needed to run the machine. During this period, there was a lack of equipment to control the pressure in the boilers, which could not withstand the pressure and ruptured, causing large numbers of accidents.

Nowadays, we have a wide source of technology, we come across new trends all the time and the speed with which everything is being developed is getting faster and faster. Plato, in his own time (300 BC), already affirmed man's dependence on technological means to survive. The continuous development of these means has allowed human beings to go to the moon, multiply, dominate - and destroy - nature.

Faced with so many forms of innovation available, sustainable awareness must be paramount in all aspects.

4.2 BRAZIL'S ENERGY SECTOR

In Brazil there are three main types of electricity generating plants: hydroelectric, conventional thermoelectric and thermonuclear, which utilise the energy contained in atomic minerals.

The share of diesel and fuel oil thermoelectric plants, the most expensive in the system, in the country's electricity production has grown year on year. The weight of these thermal plants in generation has increased by almost 286 per cent in just two years, according to data released by the National Electric Energy Agency (Aneel) on the Brazilian electricity sector in March 2014. Given this data, it is worth investing in alternative energy sources in order to reduce excessive spending and the enormous pollution caused. The hydrocarbons present in fossil fuels release greenhouse gases that can damage the ozone layer. In addition, these fuels release other gases responsible for acid rain. An example of the consequences of acid rain in Brazil was observed. The coastal municipality of Cubatão, in São Paulo, has a high concentration of industries and the acid rain has destroyed the vegetation on the slopes of the Serra do Mar, exposing the soil to erosion.

According to Figures 1 and 2, most of Brazil and the world use non-renewable sources of energy as their main sources of energy, with oil being the highlight of the world's energy matrix, causing strong environmental and social impacts.

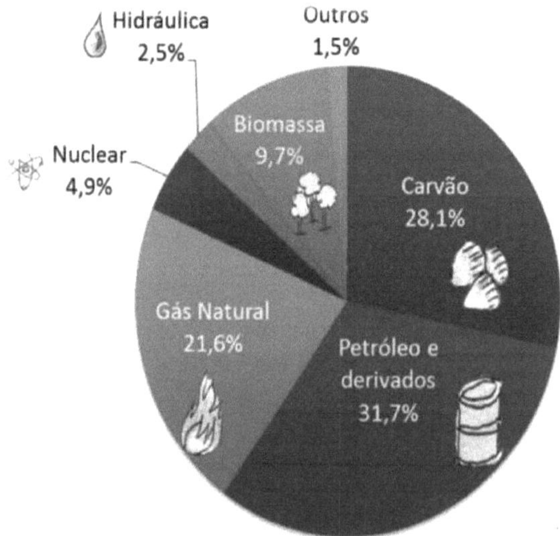

Figure 1 - World Energy Matrix

Source: http://www.epe.gov.br

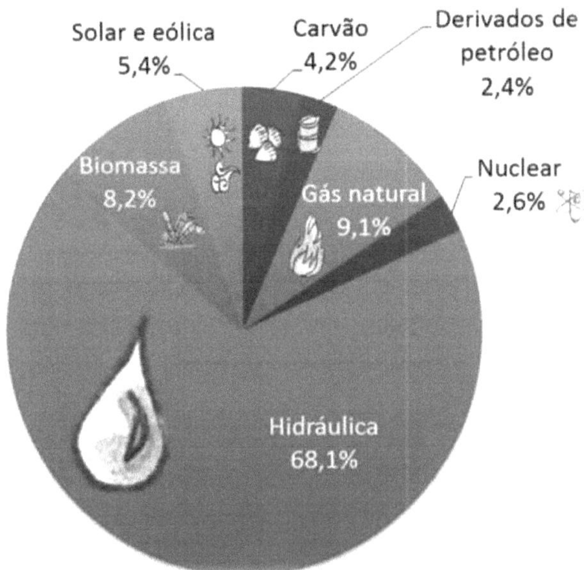

Figure 2 - Brazilian Energy Matrix
Source: http://www.epe.gov.br

4.3 FOSSIL FUELS AND THEIR ENVIRONMENTAL IMPACTS

The environment is fundamental to the life of every human being, our bodies depend on healthy nature to survive, and fuel affects this pillar to a certain extent.

> "With regard to non-renewable sources, oil reserves are expected to last for approximately 75 years, natural gas for approximately 100 years and coal for approximately 200 years"
>
> (MATTOZO, 2001).

Due to the high efficiency of these substances, they are widely used, but they are exhaustible and non-renewable sources.

According to geologists, oil, coal and natural gases are due to run out, and are therefore harmful to the environment. Although many sources are used, oil remains the main one, as many books show.

> "However, the main primary source of energy used today is still oil, which contributed

In this context, the search for ways to make it feasible to reduce the consumption of these fuels through alternative sources has been of great importance.

According to Pereira et al. (2006) the use of solar energy is advantageous throughout the country, even in regions less favoured by solar irradiation. Based on this principle, the aim is to develop a Stirling engine in which the heat source comes from sunlight.

4.4 PARABOLIC CONCENTRATOR

> The purpose of solar concentrators is to deflect the sun's rays through the reflective surface used and to focus the radiation on a reduced area, making it possible to achieve high temperatures.Solar collectors are heat exchangers that transform solar radiation into heat. The collector captures solar radiation, converts it into heat, and transfers this heat to a fluid (air, water or oil in general).
>
> (KALOGIROU, 2009)

In this project, the collector model chosen was the parabolic, which tends to concentrate all the radiation at the focus of the parabola. Although it has the highest concentration rates and is therefore the most efficient collector, it is a technology that is little used due to its high cost.

The geometry of this type of collector is fixed by two parameters, the aperture diameter D and the focal length F, or the aperture diameter D and the aperture ratio N = D/F. The figure below shows the cross-section of a circular parabolic concentrator.

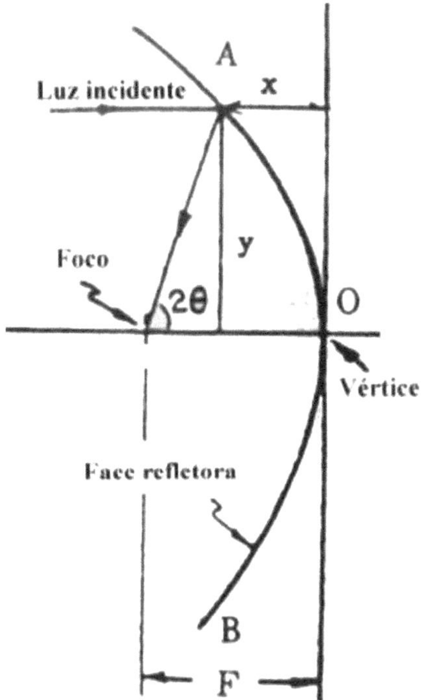

Figure 17 - Geometric diagram of a circular parabolic concentrator

Source: (Fujii,1990)

As parabolic trough collectors are the most efficient, they will be ideal in this proposed project as they will be able to transmit a higher rate of heat from irradiation to the motor, which the higher the rate received, the greater the motor's efficiency and consequently the greater the generation of electricity.

The focal length of the motor for this parabolic concentrator will be calculated and, as shown above, is directly linked to the aperture diameter of the antenna, the larger the diameter the greater the focal length, so it is not possible to define an exact focal length.

4.5 CONCENTRATION RATE

The concentration ratio is the ratio of the collector's opening area. In the case of this project, the parabolic antenna over the receiver's absorption area would be the hot chamber. It is worth emphasising that the higher the concentration ratio, the greater the heat produced and, consequently, the greater the efficiency of the motor and, with it, the greater the generation of electrical energy.

4.6 SOLAR RADIATION AND COLLECTOR TEMPERATURE

Solar irradiance is given in watts per square metre. Unlike solar radiation, which is the radiant energy emitted by the sun, it can be defined as the power of sunlight, i.e. the amount of radiation per square metre.

This irradiation is measured in instantaneous values and varies according to the time of day, place and month. It does not have the same efficiency during all the months of the year. In the north-east of Brazil, solar irradiation is higher than in the south of the country, and in the summer months this irradiation is also higher, thus increasing energy generation.

Figure 18 below illustrates the movement of the sun throughout the year. You can see that the incidence of the sun is different at different times of the year and in different parts of the world. Where June would be summer in the northern hemisphere, it would be winter in the southern hemisphere. In summer, the incidence of the sun's rays is greater and so is its irradiation.

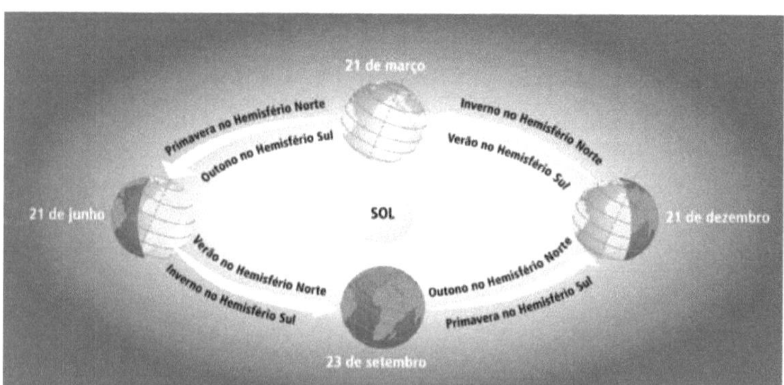

Figura 18   - Movement of the Sun throughout the year
Source: MAGNOLI, D.; SCALZARETTO. R. Geography, space, culture and citizenship.
São Paulo: Moderna, 1998. v. 1 (adapted)

It is worth pointing out that in order to know the power generated, it is necessary to know this irradiation:

$$P = A \times I$$

16

P - Power

A - Concentrator area

I - Irradiation

It can be seen that the greater this integrated irradiance over a given amount of time, the greater the power and consequently the generation of electricity.

Below is the average irradiation of Brazil compared to Europe, where it can be seen that in the former it is much higher, illustrated in Figure 19.

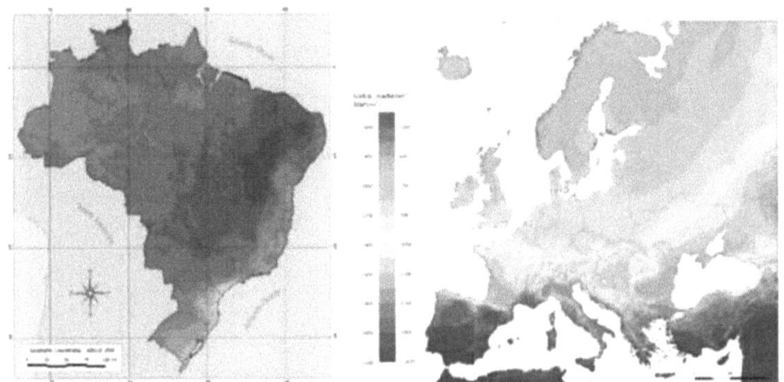

Figura 19   - Average irradiation in Brazil and Europe

Source: Average global irradiation in Brazil and Europe. Source: Brazilian Solar Energy Atlas, 2006 and PVGIS, 2012.

According to research carried out by PVGIS in 2012, Germany uses 32 per cent of the world's capacity, making it the country that generates the most electricity using solar technology.

Below is an illustration of the average annual sunshine in Brazil in terms of hours.

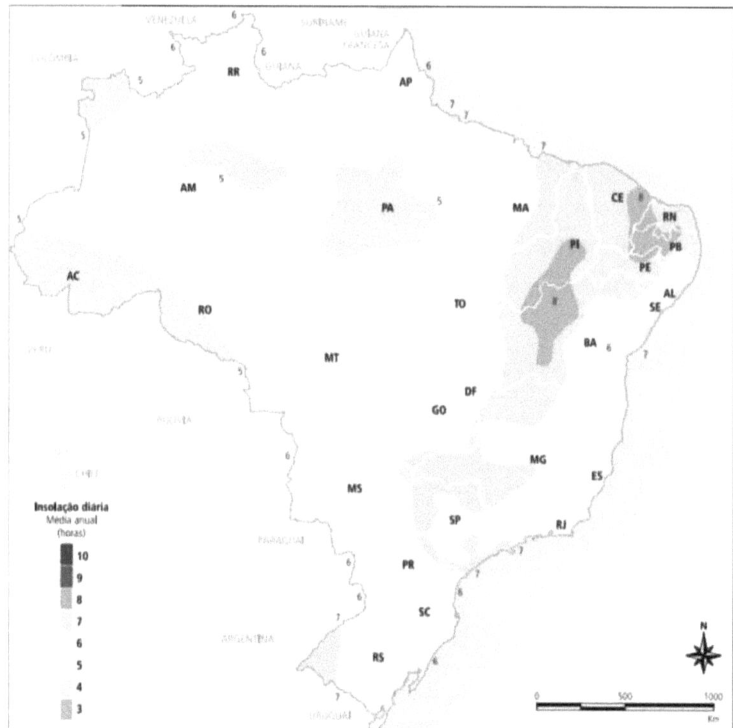

Figura 20   - Average sunshine hours

Source: ATLAS Solarímétrico do Brasil. Recife: UFPE University Press, 2000.

Figure 21 illustrates the daily global solar radiation with an annual average. Note that this radiation is given in Wh/m^2 day.

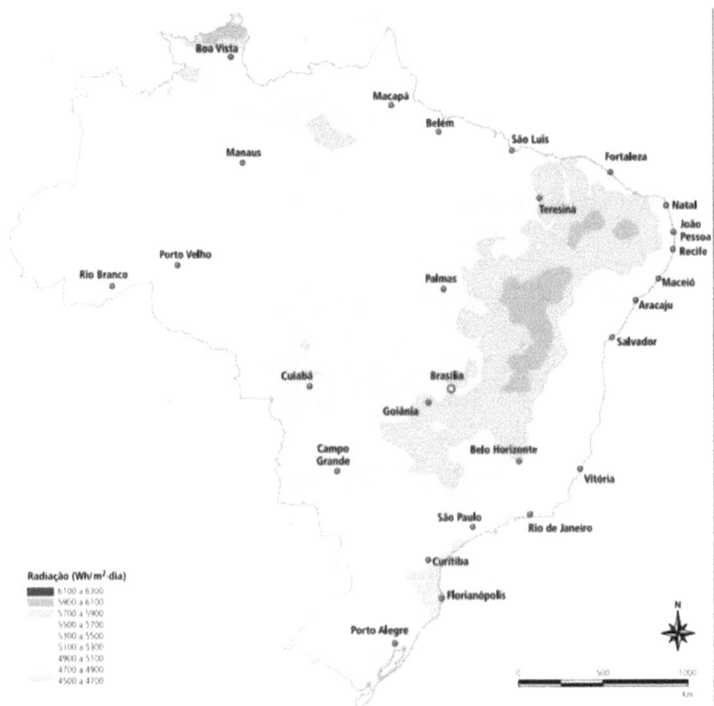

Figura 21   - Daily radiation with annual average
Source: ATLAS of Solar Irradiation in Brazil. 1998 .

CHAPTER 5

THE STIRLING ENGINE

Created by Scottish shepherd Robert Stirling and his brother in 1816, the Stirling engine was developed with the aim of replacing steam engines, which, due to the low technology of the boilers, broke down at high pressures.

The Stirling engine is an external combustion engine, which makes it even more attractive due to its flexibility to use different types of energy sources, including solar energy.

5.1 . STIRLING ENGINE MODEL ALFA

There are three types of Stirling engine: Alpha, Beta and Gamma. For the development of this project, the alpha was chosen due to its low cost, easy to understand operation and simple configuration.

The Alfa engine also needs a good seal on both pistons because it contains working gas.

> The Alfa model consists of two independent cylinders, in which the hot piston is responsible for producing the mechanical movement resulting from the engine's internal pressure and vacuum. The hot piston is located on the outside of the engine, making it visible. The two pistons together compress the working gas in the cold space, move the gas to the hot space where it expands and then returns to the cold part.
>
> (MARTINI, 1983)

Figure 2 shows the components of this engine more clearly.

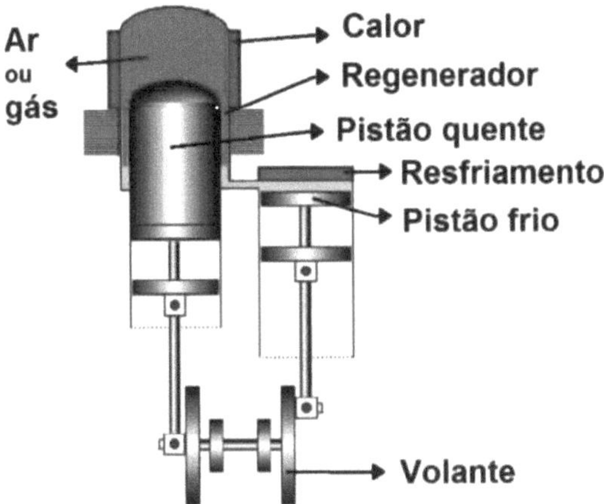

Figure 2 - Stirling engine manual
Source: Leandro Wagner - Stirling Engine Manual

5.2 DESCRIPTION OF HOW THE CYCLE WORKS

Figure 3 below shows the Stirling cycle with its isothermal compression, isochoric heating, isothermal expansion and isochoric cooling processes.

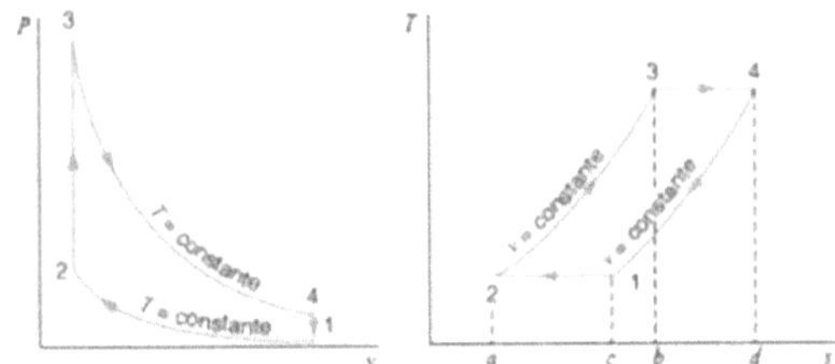

Figure 3 - Stirling cycle and its processes
Source: Van Wylen, Sonntag and Borgnakke (1995).

The best-known cycles today are Otto and Diesel, which have internal combustion. The Stirling engine, unlike those mentioned above, is external combustion, which makes it very attractive due to its flexibility to accept practically all types of fuel with an efficiency of 40%, much higher than Otto or Diesel which vary between 20% and 30%.

The Stirling cycle is made up of 4 processes classified as: compression, expansion, heating

21

and cooling, as shown in the illustrations below.

1-2. Isothermal Compression: This is when the compression and expansion pistons move upwards compressing the working fluid, while heat is rejected to the cooling system, keeping the temperature constant;

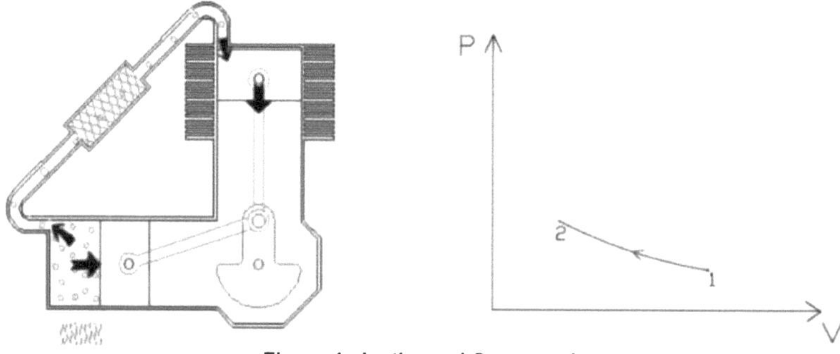

Figure 4 - Isothermal Compression
Source: Pereira, 2012.

2-3. Isochoric heating: The compression piston moves up to top dead centre, while the expansion piston moves down. The working fluid passes into the expansion space and is heated by the external source, increasing the pressure;

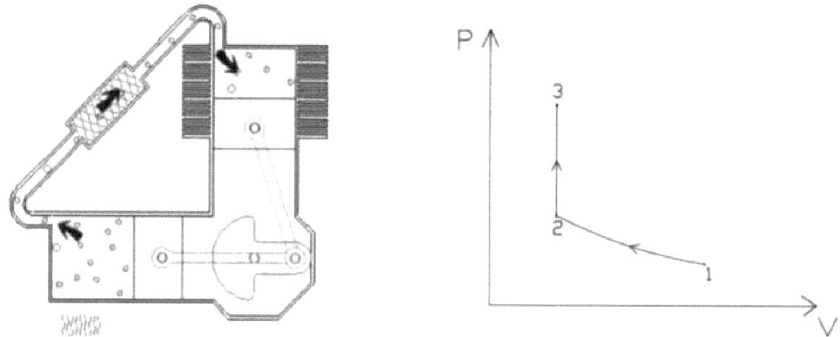

Figure 5 - Isochoric heating
Source: Pereira, 2012.

3-4. Isothermal expansion: This is the stage in which both pistons move downwards, expanding the fluid and doing work. This process takes place at constant temperature, and during expansion the working gas receives heat from the external source;

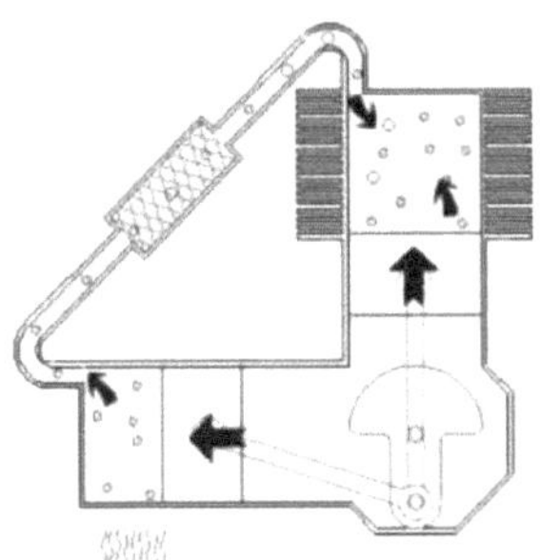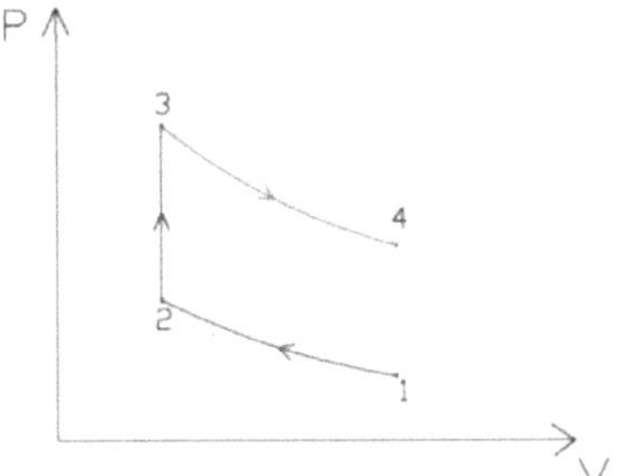

Figure 6 - Isothermal expansion

Source: Pereira, 2012.

4-1. Heat rejection at constant volume: The compression piston moves to bottom dead centre, while the expansion piston moves upwards. As a result, the working fluid passes into the compression space, having heat rejected by the cooling system, reducing its pressure to condition 1;

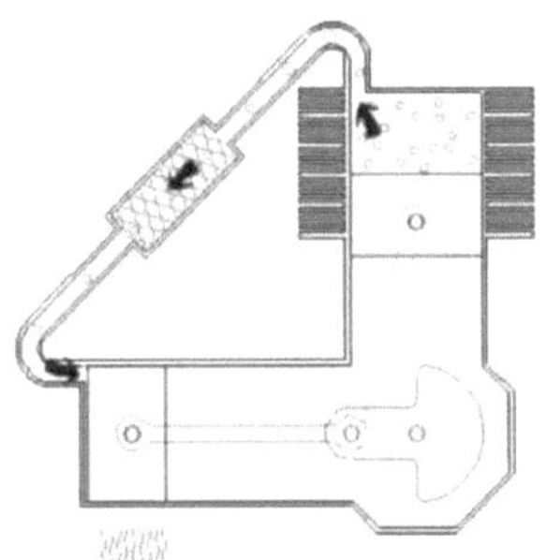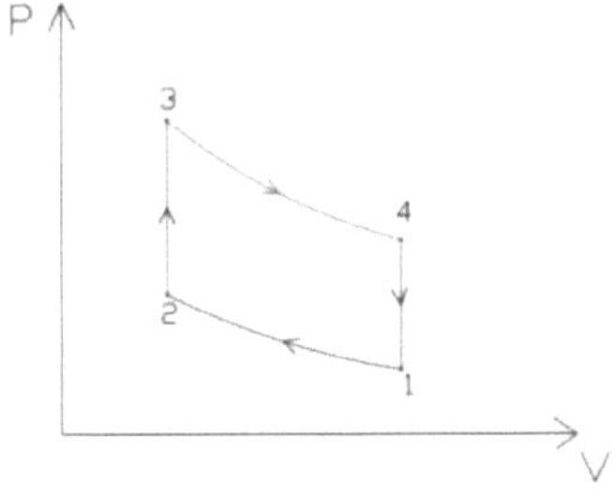

Figure 7 - Constant heat rejection
Source: Pereira, 2012

5.3 SYSTEM COMPONENTS

The Stirling engine has undergone considerable improvements in material composition and other factors over the years. It is an easy-to-understand engine and has the capacity to achieve highly considerable efficiency, making it even more attractive.

Below you can analyse the main components of this engine in isolation to better understand its functionality.

5.3.1 Heat Source

The heat source can be generated in different ways, but it is important that it does not mix with the working fluid.

5.3.2 Hot portion changer

It consists of the walls of the hot portion. The design of heat exchangers for Striling engines consists of a balance between high heat transfer, low losses and a small internal volume outside the piston stroke, the dead volume. It is important to use materials that are highly resistant to high temperatures and do not oxidise or crack.

5.3.3 Regenerating

The regenerator absorbs some of the heat that would otherwise be rejected by the engine and heats the cold air before it enters the hot chamber. Its basic operation resembles a battery, but it is a thermal accumulator. One of the benefits of the regenerator is that it improves the system's thermal efficiency.

5.3.4 Chiller and cold source

In engines with small power levels, the cooler consists of the walls of the cold portion.

It can do compressive work by cooling the volume inside itself, which is achieved through a fluid cooling system that can be natural or forced convection (DARLINGTON; STRONG, 2005, apud PRESENDO; KLIMMEK, 2009).

The cold source basically consists of the ambient temperature.

5.3.5 Steering wheel

The flywheel plays the role of keeping the engine's rotational movement constant and also absorbs vibrations.

5.4   THE BENEFITS OF THE STIRLING ENGINE

The stirling engine has great advantages, since it only needs a heat source, it is an external combustion engine and its efficiency is similar to Carnot's, i.e. even using solar energy as fuel its efficiency is largely unaffected, and it is also very quiet.

> "Another important motivation in the revival of Stirling engines is their ability to convert thermal energy into mechanical energy without explosion, as in the case of the internal combustion engine, leading to quiet and clean operation, which are essential for special applications such as military operations and medical uses"
> (TLILI, TIMOUMI, NASRALLAH, 2006).

Another benefit of the Stirling engine is that, as it is an external combustion engine, it is possible to control the emission of polluting gases and thereby reduce them. With the advance of technology, it is possible to seek sustainable development, to evolve without harming the environment, which is what all large companies and universities are striving for.

As the Stirling engine only needs one source of heat and does not depend on the type of fuel, it can not only work but also be highly efficient using solar energy, which is why this engine is being studied more and more, Interest in this engine has been revived with the intention of minimising the use of fossil fuels and using more alternative fuels, thus reducing the emission of polluting gases and thus being advantageous to the environment and consequently to human beings, and is also due to advances in material technology, manufacturing processes and sealing systems, factors that in the past made the Stirling engine unviable. (BARROS, 2005)

CHAPTER 6

SCHMIDT'S THEORY

6.1   GENERAL CONSIDERATIONS

In 1871 Gustav Schmidt developed the first analysis of Stirling engines. Known as Schmidt's theory, it is the simplest and most widely used when it comes to initial engine development. It can be very useful for engine design as it is certainly more realistic than the ideal Stirling cycle, which is why it remains idealised. It is based on the expansion and compression of an ideal gas (WALKER, 1980).

The performance of a Stirling engine is estimated using a P-V diagram. The engine volume is calculated from the geometry of the engine being analysed. This estimate is made using the ideal gas equation and method, as shown in the equation:

$$PV = mRT$$

(1)

In which:

P = pressure

V = volume

m = mass

R = universal gas constant

T = temperature

In order to carry out the analysis, some preliminary considerations must be made, as presented by Sulzbach (2010).

a)   There is no loss of pressure during heat exchange and there are no different internal pressures.

b)   The process of expansion and compression takes place isothermally.

c)   The working gas must be considered ideal.

d)   There is a perfect regenerator.

e)   The expansion dead volume maintains the temperature of the expanding gas (TE), the compression dead volume maintains the temperature of the compressing gas (TC).

f)   The regenerator temperature is an arithmetic mean of the temperatures of the expansion (TE) and compression (TC) cylinders.

g)   Expansion volume (EV) and compression volume (CV) vary according to a sine curve.

Table 1 shows the variables of Schmidt's theory and their units of measurement respectively. The equations for the calculation are similar for the engine regardless of whether it is alpha, beta or gamma.

Table 1 - Schmidt theory variable

Description	Symbol	Unit
Engine pressure	P	Pa
Swept volume of the expansion or displacement piston	V_{SE}	m^3
Swept volume of the compression piston or working piston	V_{SC}	m^3
Expansion space dead volume	Kr>r	m^3
Regenerator volume	V_{DE}	m^3
Dead volume of the compression space	V_{DC}	m^3
Momentary volume of expansion space	V_E	m^3
Momentary volume of the compression space	You	m^3
Total momentary volume	V	m^3
Total mass of the working fluid	m	kg
Gas constant	R	$J\,kqK$
Gas temperature in the expansion space	T_H	K
Gas temperature in the compression space	T_G	K
Gas temperature in the regenerator space	T_R	K
Phase angle	dx	9
Temperature ratio	t	dimensionless
Swept volume ratio	v	dimensionless
Dead volume ratio	X	dimensionless
Motor speed	n	Hz
Expansion energy	W_E	J
Compression energy	W_c	J
Net energy	W_J	J
Expansion power	L_E	W
Compression power	Lc	W
Net power	Li	W
Efficiency	e	dimensionless

Source: Hirata (1997) apud Presendo; Klimmek (2009)

6.2 SCHMIDT EQUATIONS FOR THE ALPHA-TYPE STIRLING ENGINE

Figure 8 below shows the Schmidt equations for the alpha-type Stirling engine.

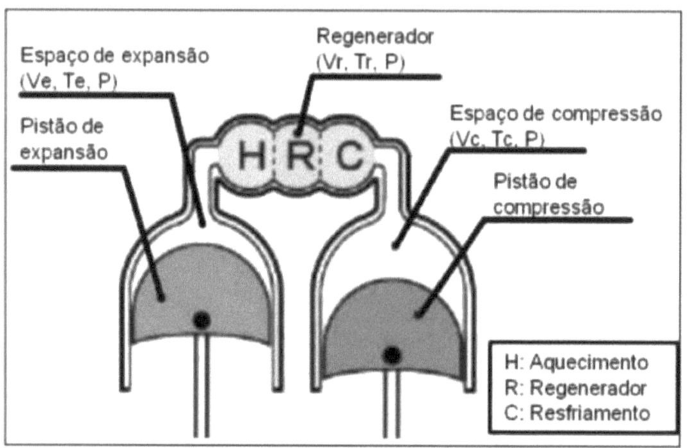

Figure 8 - Schmidt equations for the Alpha-type Stirling engine
Source: Hirata, 1995.

Firstly, it is necessary to determine the expansion and compression volumes for a given engine rotation angle. The instantaneous volume is described by an x-axis angle. This axis angle is defined as x = 0 when the expansion piston is located in the top position (top dead centre). The instantaneous expansion volume - VE is described by eq. 2 as a function of the volume travelled by the expansion piston - VSE, a dead expansion volume - VDE under the assumed conditions (g).

$$V_E = \frac{V_{SE}}{2}(1 - \cos x) + V_{DE}$$

(2)

The momentary compression volume is found using Equation 3 as a function of the volume travelled by the compression piston, the compression dead volume and the phase angle.

$$V_C = \frac{V_{SC}}{2}[1 - \cos(x - dx)] + V_{DC}$$

(3)

The total momentary volume is calculated using Equation 4

$$V = V_E + V_R + V_C$$

(4)

Using the assumptions in (a), (b) and (c), the total mass (m) of gas contained in the engine is calculated as a function of pressure (P), each temperature (T), each volume (V), and the gas constant (R).

$$m = \frac{PV_E}{RT_E} + \frac{PV_R}{RT_R} + \frac{PV_C}{RT_C}$$

(5)

The relative temperature (t), relative travelled volume (v) and relative dead volumes are found in the following equations.

$$t = \frac{T_C}{T_E}$$

(6)

$$v = \frac{V_{SC}}{V_{SE}}$$

(7)

$$X_{DE} = \frac{V_{DE}}{V_{SE}}$$

(8)

$$X_{DC} = \frac{V_{DC}}{V_{SE}}$$

(9)

$$X_R = \frac{V_R}{V_{SE}}$$

(10)

The regenerator temperature is calculated by Equation 11 using assumption (f).

$$T_R = \frac{T_E + T_C}{2}$$

When equation (5) is modified using equations (6) and (11), the total mass of gas contained in the engine is described by Equation 12.

$$m = \frac{P}{RT_C}\left(tV_E + \frac{2\tau V_R}{1+t} + V_C\right)$$

$$(12)$$

Equation 12 is modified using Equations 2 and 3, and the total mass of gases (m) is described by Equation (13).

$$m = \frac{PV_{SE}}{2RT_C}\{S - Bcos(x - a)\}$$

$$(13)$$

Now,

$$a = tg^{-1}\frac{v.\,sen\,dx}{t + cos\,dx}$$

$$(14)$$

$$S = t + 2tX_{DE} + \frac{4tX_R}{1+t} + v + 2X_{DC}$$

$$(15)$$

$$B = \sqrt{t^2 + 2tvcos\,dx + v^2}$$

$$(16)$$

The engine pressure (P) is defined by Equation 17, using Equation 13:

$$P = \frac{2mRT_C}{V_{SE}\{S - Bcos(\theta - a)\}}$$

$$(17)$$

The average pressure is calculated below:

$$P_{med} = \frac{1}{2\pi}\int P dx = \frac{2mRT_C}{V_{SE}\sqrt{S^2 - B^2}}$$

$$(18)$$

is defined by the following equation.

$$c = \frac{B}{S}$$

(19)

As a result, the engine pressure (P), as a function of the average pressure, is calculated in Equation 20.

$$P = \frac{P_{med}\sqrt{S^2 - B^2}}{S - B\cos(x - a)} = \frac{P_{med}\sqrt{1 - c^2}}{1 - c.\cos(x - a)}$$

(20)

On the other hand, when cos(x - a) = -1, the engine pressure (P) becomes minimal. The next equation is introduced:

$$P_{min} = \frac{2mRT_C}{V_{SE}(S + B)}$$

(21)

Therefore, the engine pressure (P), as a function of the minimum pressure, is described in the following equation.

$$P = \frac{P_{min}(S + B)}{S - B\cos(x - a)} = \frac{P_{min}(1 + c)}{1 - c.\cos(x - a)}$$

(22)

Similarly, when cos(x - a) = 1, the engine pressure (P) becomes maximum. Equation 23 is formulated.

$$P = \frac{P_{max}(S - B)}{S - B\cos(x - a)} = \frac{P_{max}(1 - c)}{1 - c.\cos(x - a)}$$

(23)

CHAPTER 7

ENGINE DESIGN AND STUDY

In this topic we will discuss the proposal for a Stirling engine coupled to a parabolic dish and how it works in general, describing the materials chosen for the development so that it achieves good efficiency levels.

7.1   THE SOLAR-POWERED STIRLING ENGINE PROPOSAL

The project involves the study of a Stirling engine coupled to a parabolic antenna. The antenna is covered with a mirrored strip and its purpose is to capture solar energy and concentrate it in a specific way to run the engine. It is worth emphasising the importance of using this heat source because of its environmental benefits and because it can be transformed into electrical energy.

The Stirling engine operates in four strokes, which can be categorised as compression, expansion, heating and cooling. This engine is considered simple because it uses only two chambers with different temperatures to produce work, and unlike the Otto or Diesel engines, it can use any type of fuel, making it ideal for the proposed work.

This engine works with the four strokes mentioned above. For the engine to work, there must be a difference in the temperatures of the chambers.

The solar energy will be captured by means of a parabolic antenna covered with a mirrored ribbon, since as it is internally mirrored to form a "concave mirror" it causes the sun's rays to fall on a single point, emphasising that this collector will be facing the engine at a focal distance previously calculated according to the diameter of the antenna, generating a hot vapour.This gas enters the hot chamber, carrying out the heating and expansion process. As a result, the gas expands and ends up being transferred by the displacement piston, forcing the working piston upwards into the cold chamber, where it cools down and ends up becoming a vapour.

contracting, i.e. decreasing its volume, forcing the piston downwards, where the cycle restarts.

According to the company MyBeloJardim, Arizona-USA was the first commercial plant in the world to use Stirling technology to concentrate solar energy in giant parabolic discs to generate electricity from the sun. 60 solar parabolic collectors, called SunCatchers, were installed at the solar plant. Each parabolic unit is capable of producing 25 kilowatts and the entire installation has a capacity of 1.5 megawatts of electricity generation. Enough to supply the energy needs of around 200 homes in January 2010.

Figure 9 - Stirling Solar Plant - Arizona USA
Source: http://mybelojardim.com/inaugurada-usina-solar-sterling-de-geracao-de- electricity/

7.   2ENGINE DEVELOPMENT

At this stage, it is possible to analyse the parameters for an experimental project. The alpha-type Stirling engine is simple to build because it is driven by two pistons.

Below you can analyse the projections of the engine's main components with their appropriate composition in Figure 10.

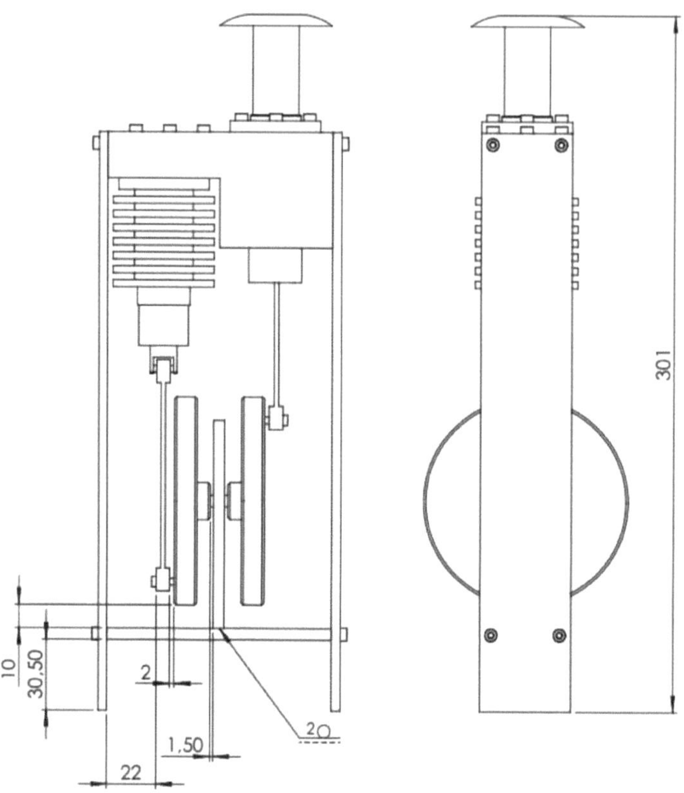

Figure 10-Project engine

Source: Author

Figure 11 shows a functional analysis of all the engine components, designed in SolidWorks.

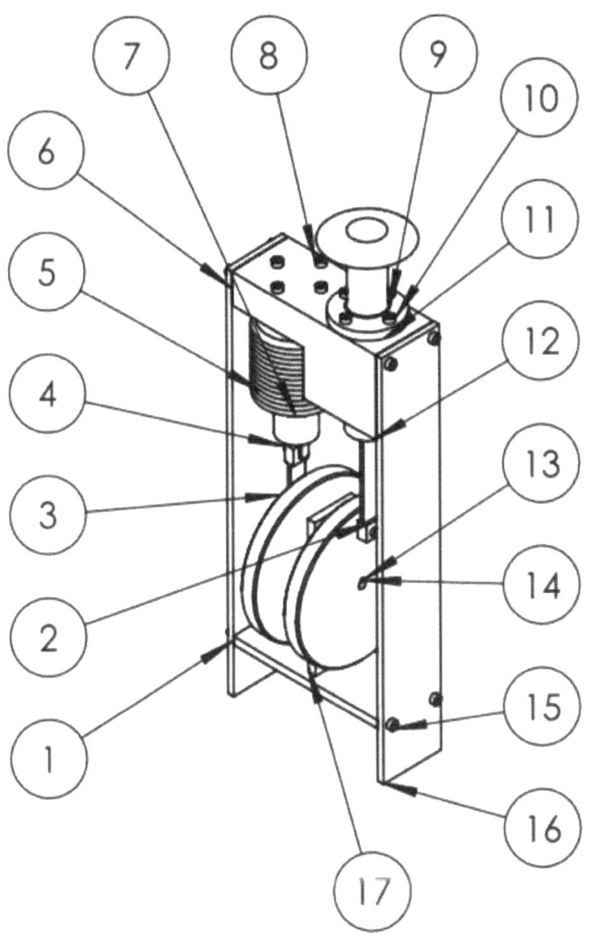

Figure 11 - Engine and its components
Source: Author

N°	Qty.	Name	Material / Specification
1	1	Lower base	Aluminium
2	2	Connecting rod	Aluminium
3	2	Steering wheel	Aluminium
4	1	Plunger	Aluminium
5	1	Cold housing	Aluminium
6	1	Block	Aluminium
7	1	Piston cylinder	Aluminium
8	8	M3 x 12 bichromatised allen head screw	Aluminium
9	1	Plunger 2	Aluminium
10	1	Hot housing	Aluminium

11	1	Piston cylinder 2	Aluminium
12	2	Key	Aluminium
13	1	Axis	Aluminium
14	8	M3 x 12 bichromatised allen head screw	Aluminium
15	2	Sup lateral	Aluminium
16	1	Steering wheel support	Aluminium

7.2.1 Main components

Hot housing

Knowing that the Stirling engine works by the temperature difference between the hot and cold chambers, the hot housing will be facing the antenna at a previously calculated focal distance from the parabolic concentrator, it will receive the hot vapour transmitted after the sun's rays meet the concentrator. The hot housing will heat up the air and thereby expand, as shown in Figure 12.

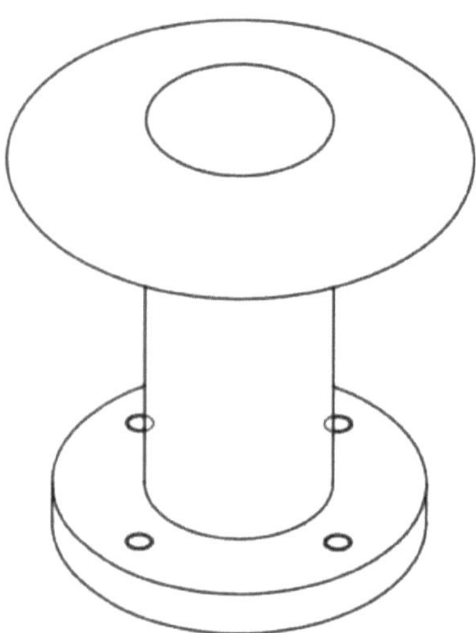

Figure 12- Hot carcass
Source: Author

Cold housing

After passing through the hot chamber, the expanded and heated air passes through a cooler with a temperature difference. This cold housing, as illustrated in Figure 13, has the function of cooling the air and then compressing it, causing the fluid to return to the beginning and thus completing the thermodynamic cycle.

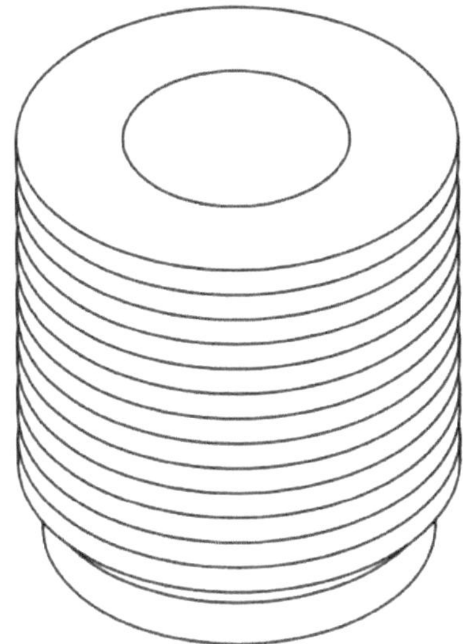

Figure 13- Cold housing
Source: Author

Connecting rod

The connecting rod, as shown in image 14, has the function of transforming linear movement into circular movement. In the engine, it is responsible for connecting the piston to the crankshaft.

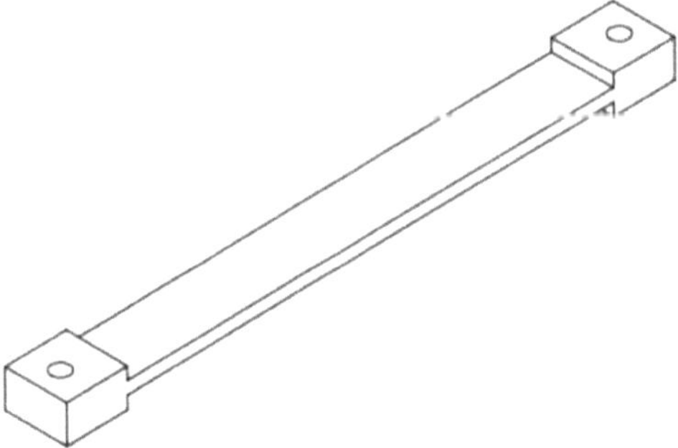

Figura 14   - Connecting rod
Source: Author

Steering wheel

The flywheel, as shown in image 15, has the function of balancing the sudden impulses of the piston without damaging the engine, thereby reducing the impacts and vibrations of the engine and avoiding possible future damage.

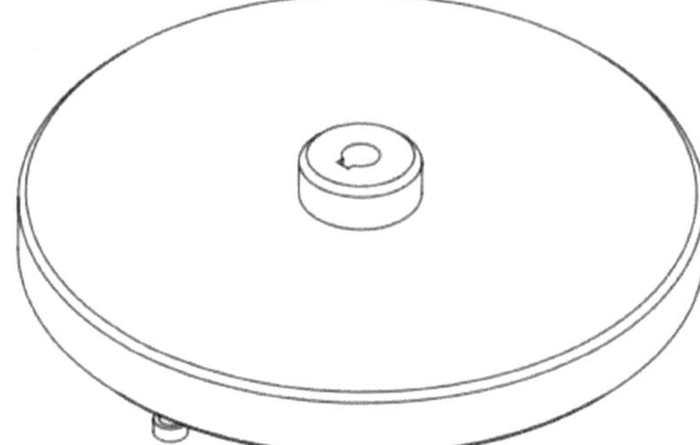

Figura 15 - Steering wheel
Source: Author

The items block, spar, flywheel support, lower base, piston base are used to connect the engine. All the projections mentioned above are detailed in the appendices at the end of the monograph.

7.   3FUNCTIONING

The engine will work as follows: the parabolic concentrator will receive the radiation and reflect it to the hot housing, which will be facing the concentrator. The air will heat up and expand, passing through the cooler where there will be a temperature difference, causing the piston to move and the engine to start moving.

Once the motor is running, it will be connected to a dynamo, which is an electric current generator, using this running motor to generate electrical energy, i.e. transforming this mechanical energy into electrical energy so that it can be coupled to batteries or connected directly to the consumer's electricity grid.

In Figure 16, we can analyse the 3D motor design developed in SolidWorks:

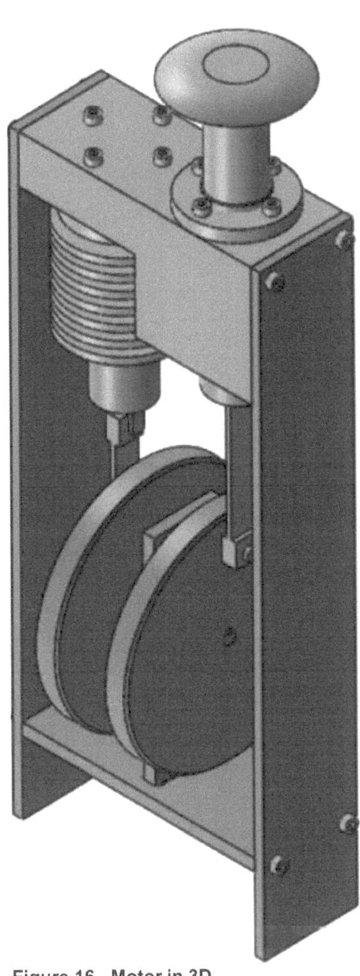

Figure 16 - Motor in 3D
Source: Author

CHAPTER 8

BUDGET

A budget analysis of the prototype was carried out, including the machining of the motor, the antenna, the dynamo and the mirrored tape so that the sun would hit the motor, as shown in the following tables:

Table 1 - Machining budget

COMPANY:	VALUE (R$)
RS metals and machining	R$ 1300,00
LG Machining	R$ 1400,00
Machining in Porto Real	R$ 1250,00

Table 2 - Antenna budget

COMPANY:	VALUE (R$)
CLEAR	R$ 144,99
SKY	R$ 122,90

Table 3 - Budget Mirrored tape

COMPANY:	VALUE (R$)
Link 7	R$ 40,18

Table 4 - Dynamo budget

COMPANY:	VALUE (R$)
Fonseca Bicycle Shop	R$ 200,00
BM Bicycle Shop	R$ 100,00

All the prices shown are inclusive of freight charges. An average quote for this engine can vary from R$1513.08 to R$1786.17 from the companies described.

CHAPTER 9

RESULTS OBTAINED

As discussed in the course of this study, irradiation varies both during the months of the year and according to location, affecting both the incidence in watts and the average insolation in hours.

After analysing the months of the year together with the annual average in Brazil, we can conclude that the average monthly irradiation is 900 Watts/m^2 .

It should be emphasised that the larger the antenna, the more power it will generate, since the radius of the antenna is directly linked to the power acquired by the motor. In this study, we designed an antenna with a diameter of 1 metre, i.e. a radius of 50 centimetres.

According to the ANEEL table, 1kW/h is around R\$0.87295, bearing in mind that the efficiency of the Stirling engine is around 45%. With the above data we can analyse the power results.

To analyse the power generated, it is necessary to multiply the area of the dish (A= 0.5*n*radius of the dish) with the average irradiation and the efficiency of the motor.

$$P = A * I * Ef$$

$$P = (0,5 * 3,1416 *0,5) *900*0,45$$

$$P = 317.93 \text{ Watts}$$

In order to analyse the value of the savings, it is necessary to transform this power into Kw/h as given in the ANEEL table.

So multiplying the power by the number of days in the month, on average thirty, and the average in hours, which according to CRESESB is 8 hours a day, shows that:

$$P * 30 * 8 = 76.30 \text{ kW/h}$$

$$317.93 * 30 * 8 = 76.30 \text{ kW/h}$$

By carrying out this transformation, it is possible to analyse the power found in kW/h and multiply it by the rate of 1kW/h given in the ANEEL table.

$$76.30 * 0.87295 = 66.60 \text{ reais}$$

You can thus find a monthly saving of R$66.60.

CHAPTER 10

CONCLUSION

This work presented and demonstrated the importance of evolution without neglecting sustainability. It looked at the operation of a Stirling engine coupled to a parabolic collector in order to generate electricity using the sun's rays.

Given that hydropower is the main source in Brazil and that water, despite being a clean source, is not renewable, it is very important to think about alternative sources that can support the generation of electricity and in the future could become a primary source. As well as being a clean source, solar energy is renewable, free of charge and free of polluting gases, thus reducing the emission of CO_2 into the atmosphere, a gas that is highly responsible for the greenhouse effect.

Specific analyses were made of the theoretical operation of the Stirling engine, which in addition to its many benefits, from the flexibility to use various fuels, to its high efficiency compared to Carnot, is an engine under study and development, but which in the very near future is likely to become affordable.

BIBLIOGRAPHICAL REFERENCES

BARROS, RobledoWakin. Theoretical and Experimental Evaluation of the Solo 161 Stirling Engine Operating with Different Fuels. Itajubá: UNIFEI, 2005. Available at: http://juno.unifei.edu.br/bim/0030363.pdf. Accessed on: 03/06/2018

BAHNEMANN Wellington and AMORIN, Daniela, Agencia Estado. ANEEL data.
Available at: https://economia.estadao.com.br/noticias/geral,peso-de- termoeletricas-cresceu-286-em-2-anos,178924e, Accessed on: 07/06/2018

CACHUTÉ, L. Theoretical analysis and presentation of a Stirling engine design methodology for use in an evaporative cooling system

CARDOSO, Mayara. Available at: https://www.infoescola.com/fisica/ciclos-termodinamicos/ accessed on 02/06/2018

CRUZ, Vinicius Guimarães da. Experimental development of a gamma-type Stirling engine. Federal University of Paraíba: João Pessoa, 2012.

DARLINGTON, R.; STRONG, K. Stirling and hot air engines: designing and building experimental model Stirling engines. Crowood: Crowood Press, 2005.

DUARTE, MichelleAvailable at:https://www.todamateria.com.br/combustiveis-fosseis/ accessed on 02/06/2018.

GARCIA, ArmandoSuarezAvailable in :
https://www.suapesquisa.com/energia/consumo_energia_brasil.htm accessed on

HIRATA, K. (1995). SchmitdtTheory for StirlingEngines. StirlingEngine Home Page. Available at <hptt://www.bekkoame.ne.jp/~khirata/>. Accessed on: 03/12/2018

KALOGIROU, S. A. Solar energy engineering: processes and systems. 1st ed. San Diego (USA): Elsevier, 2009.

KONGTRAGOOL, B.; WONGWISES, S. A reviewof solar powerredStirlingenginesandlowtemperaturedifferentialStirlingengines. RenewableandSustainable Energy Reviews v. 7, p 131- 154, 2003. Available at: http://www.inference.phy.cam.ac.uk /sustainable/refs/ solar/Stirling.pdf. Accessed on: 03/12/2018

MARTINI, W. R.; Stirling engine design-manual. Honolulu: University Press of the Pacific, 1983.

PEREIRA, E. B; Martins, F. R.; Abreu, S. L.; Ruther, R. 2006. Brazilian solar energy atlas. National Institute for Space Research (INPE), São José dos Campos, 60p

PEREIRA, Felipa. Classification of Stirling Engines According to Configuration. Accessed on 10 September 2018.

SULZBACH, Jaimir. Design and manufacture of a didactic model stirling engine. Panambi: UNIJUI, 2010.

TLILI, I., TIMOUMI, Y., NASRALLAH, S. B., Analysisand design considerationofmeantemperaturedifferentialStirlingengine for solar application. Renewable Energy, v. 33, p. 1911-1921, 2008.

VAN WYLEN, G.; SONNTAG, R. E.; BORGNAKKE, C. Fundamentals of classical thermodynamics. Translation of the 4th American edition. São Paulo: Editora Edgard Blucher Ltda, 1995.

WALKER, Graham. Stirling engines. Oxford: Oxford University Press. 1980 (Oxford Science Publications).

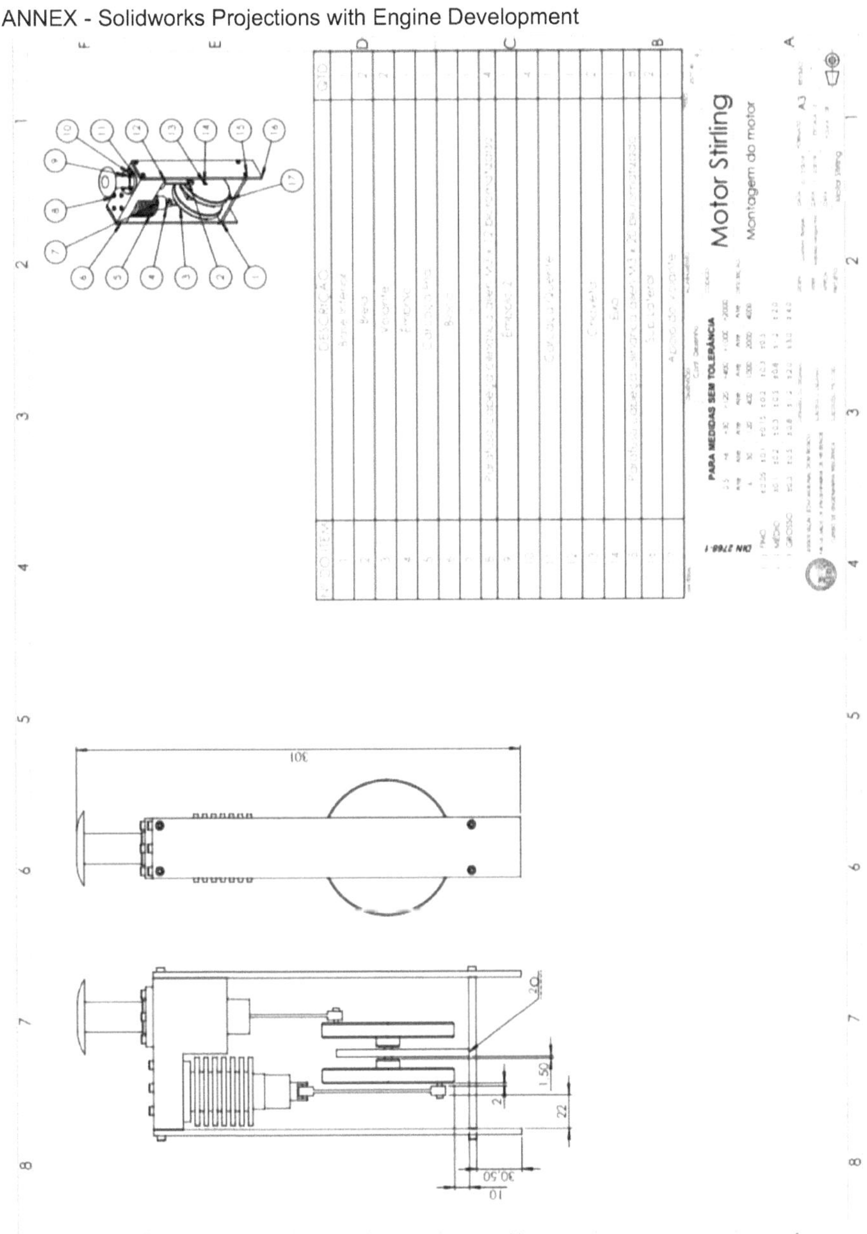

Motor Stirling
Montagem do motor
DESCRIÇÃO
PARA MEDIDAS SEM TOLERÂNCIA
DIN 2768-1

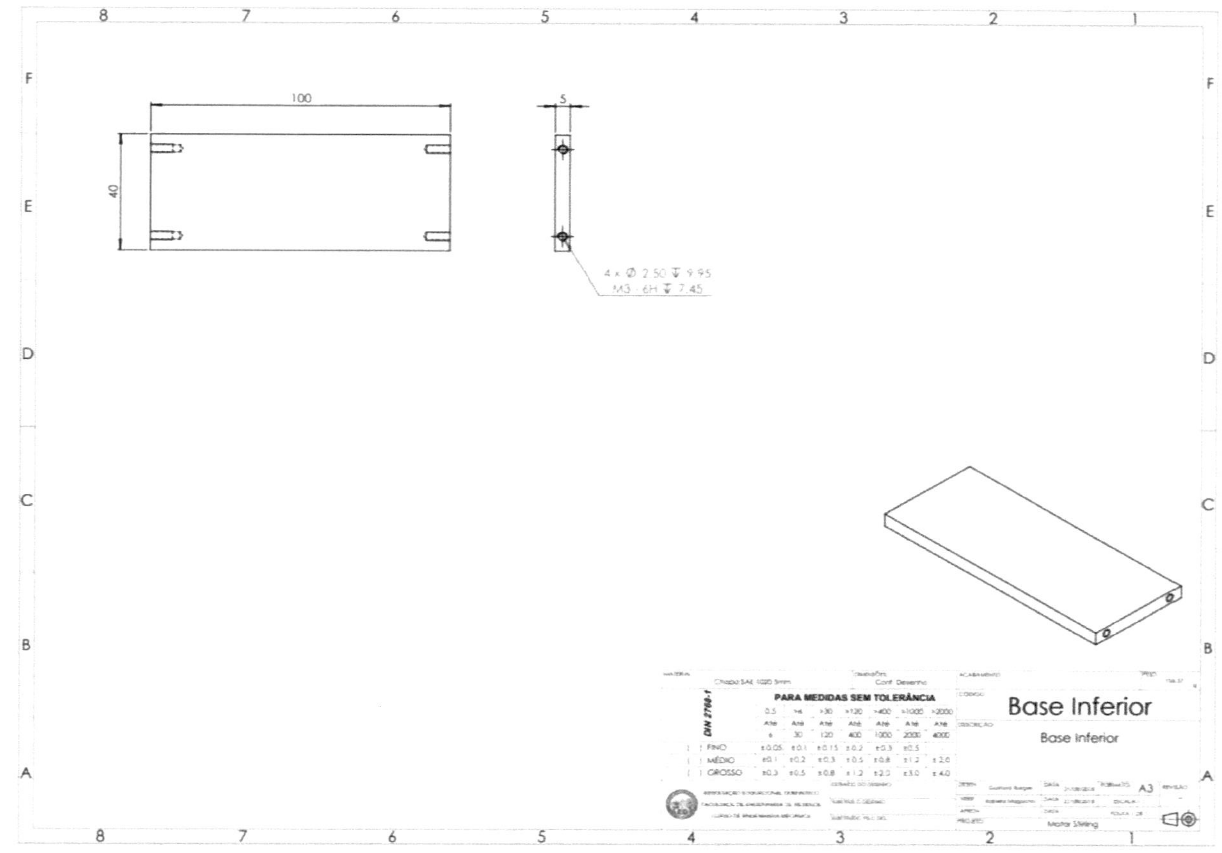

100
40
5
4 x Ø 2.50 ▼ 9.95
M3 - 6H ▼ 7.45
Chapa SAE 1020 5mm
Conf. Desenho
PARA MEDIDAS SEM TOLERÂNCIA
DIN 2768-1
0.5 | >6 | >30 | >120 | >400 | >1000 | >2000
Até | Até | Até | Até | Até | Até | Até
6 | 30 | 120 | 400 | 1000 | 2000 | 4000
FINO | ±0.05 | ±0.1 | ±0.15 | ±0.2 | ±0.3 | ±0.5
MÉDIO | ±0.1 | ±0.2 | ±0.3 | ±0.5 | ±0.8 | ±1.2 | ±2.0
GROSSO | ±0.3 | ±0.5 | ±0.8 | ±1.2 | ±2.0 | ±3.0 | ±4.0
Base Inferior
Base Inferior
A3
Marta Stirling

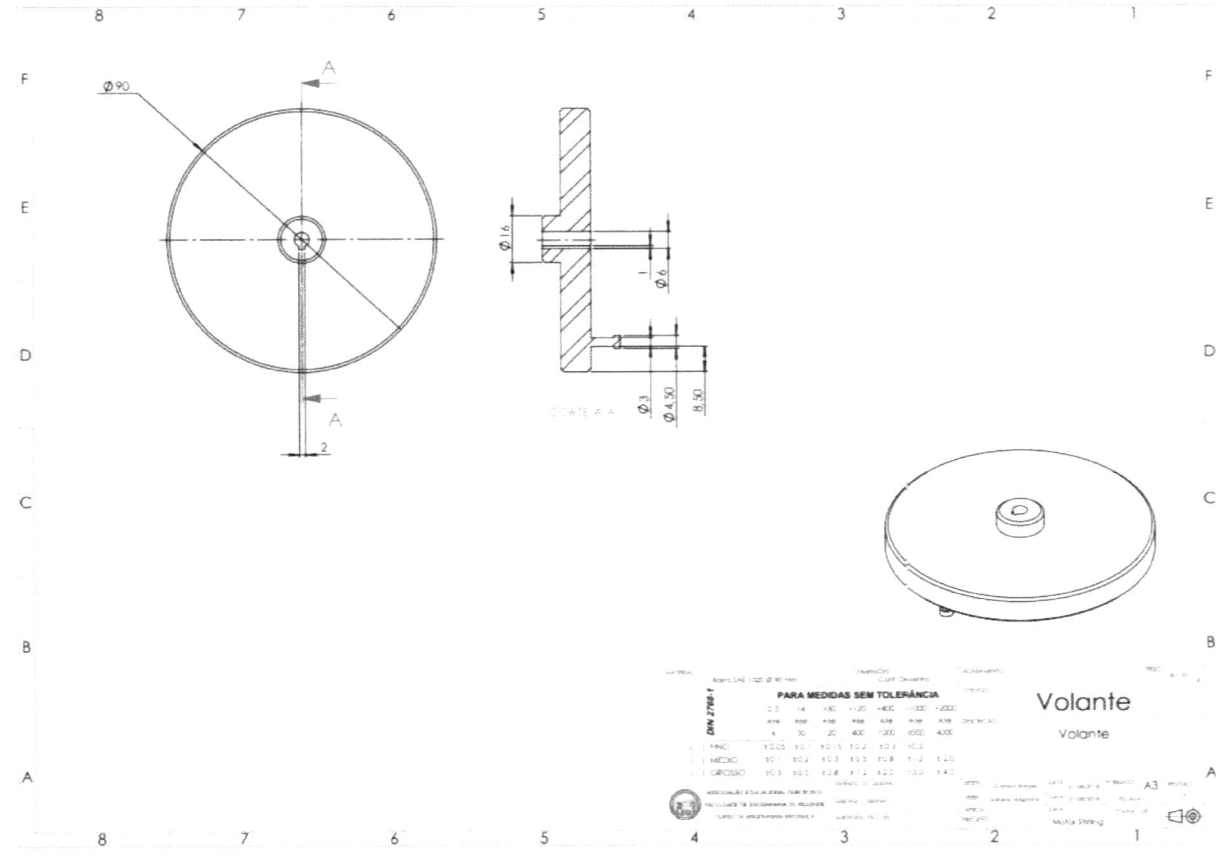

Ø90
Ø16
Ø3
Ø4,50
8,50
Ø6
DIN 2768-1
PARA MEDIDAS SEM TOLERÂNCIA
Volante
Volante
A3

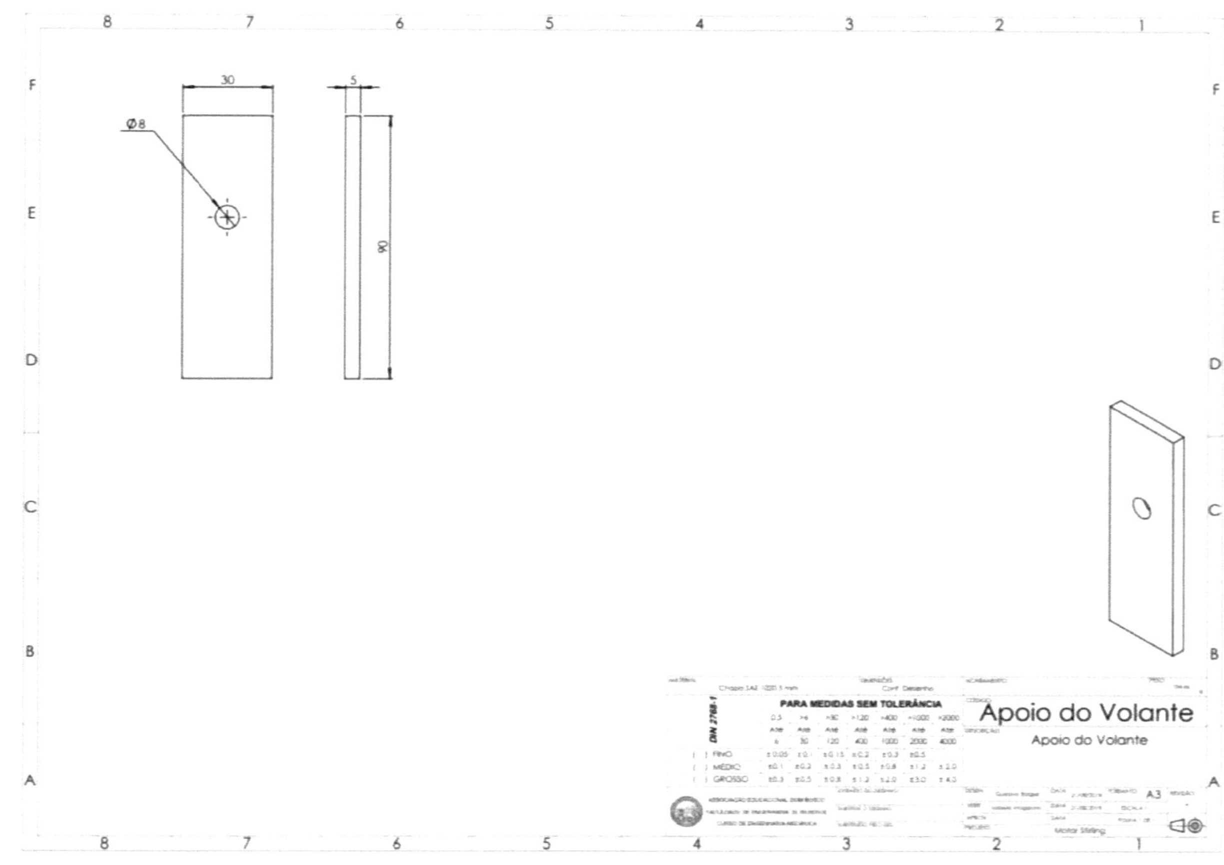

Ø8
30
5
90
PARA MEDIDAS SEM TOLERÂNCIA
DIN 2768-1
Apoio do Volante
Apoio do Volante
A3
Motor Stirling

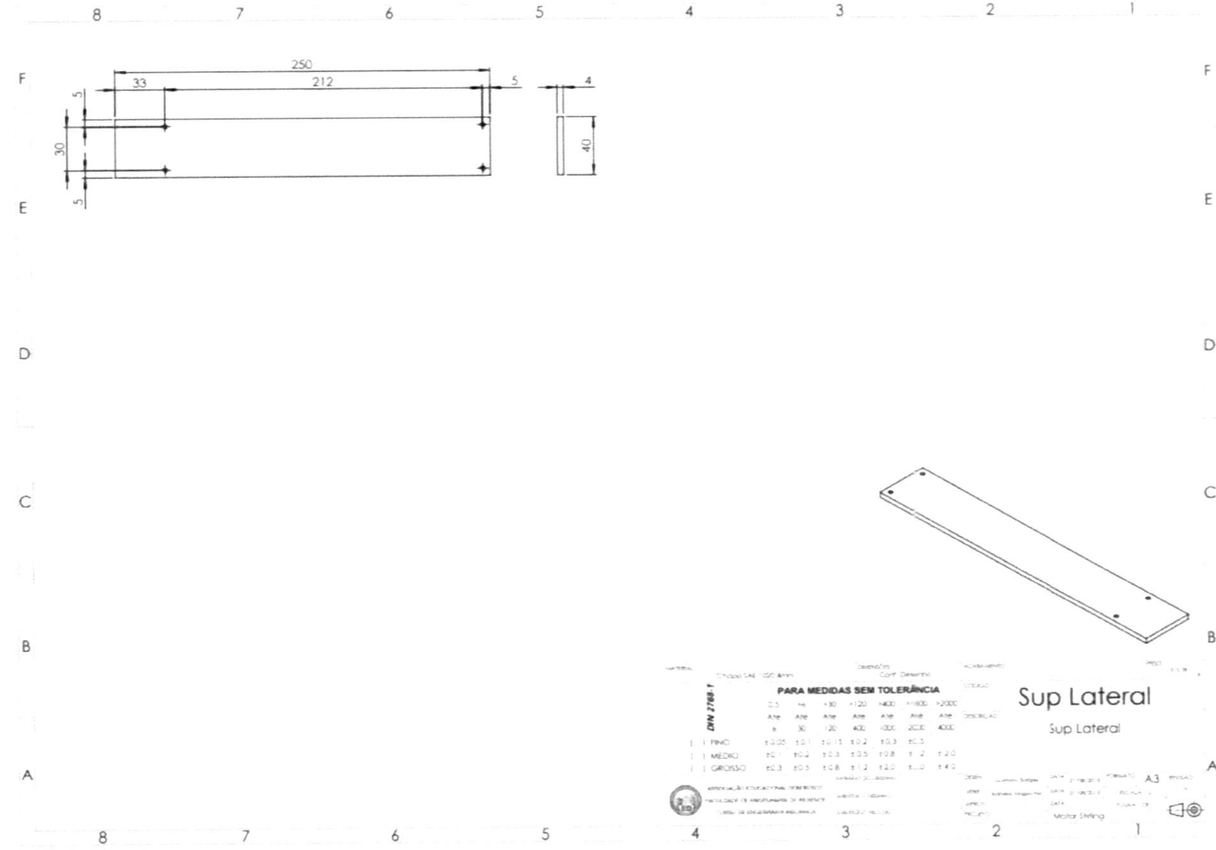

250
212
33
30
5
5
5
40
4
PARA MEDIDAS SEM TOLERÂNCIA
DIN 2768-1
Chapa SAE 1020 4mm
FINO
MEDIO
GROSSO
Sup Lateral
Sup Lateral
A3
Motor Stirling

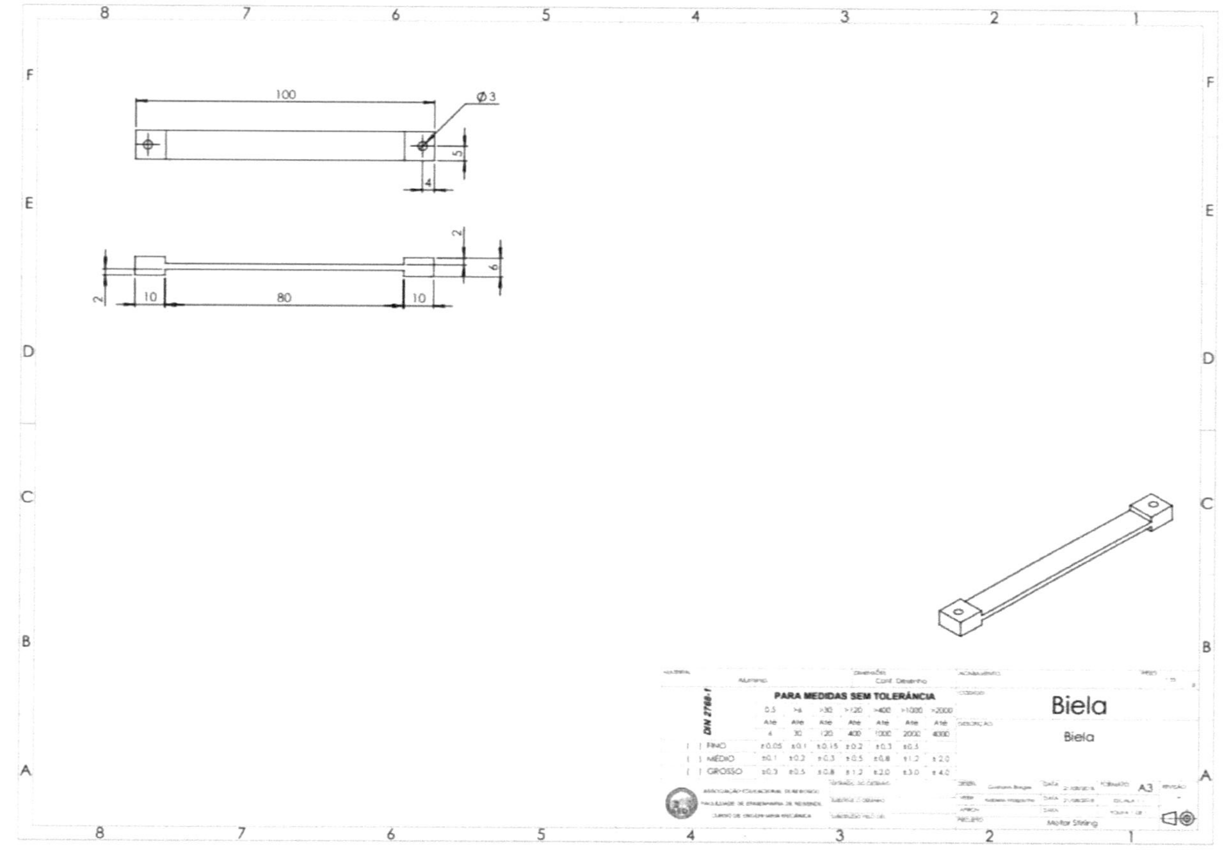
Ø3
100
5
4
2
6
2
10
80
10
PARA MEDIDAS SEM TOLERÂNCIA
DIN 2768-1
0.5 >6 >30 >120 >400 >1000 >2000
Até Até Até Até Até Até Até
6 30 120 400 1000 2000 4000
FINO ±0.05 ±0.1 ±0.15 ±0.2 ±0.3 ±0.5
MÉDIO ±0.1 ±0.2 ±0.3 ±0.5 ±0.8 ±1.2 ±2.0
GROSSO ±0.3 ±0.5 ±0.8 ±1.2 ±2.0 ±3.0 ±4.0
ASSOCIAÇÃO EDUCACIONAL DOM BOSCO
FACULDADE DE ENGENHARIA DE RESENDE
CURSO DE ENGENHARIA MECÂNICA
Biela
Biela
A3
Motor Stirling

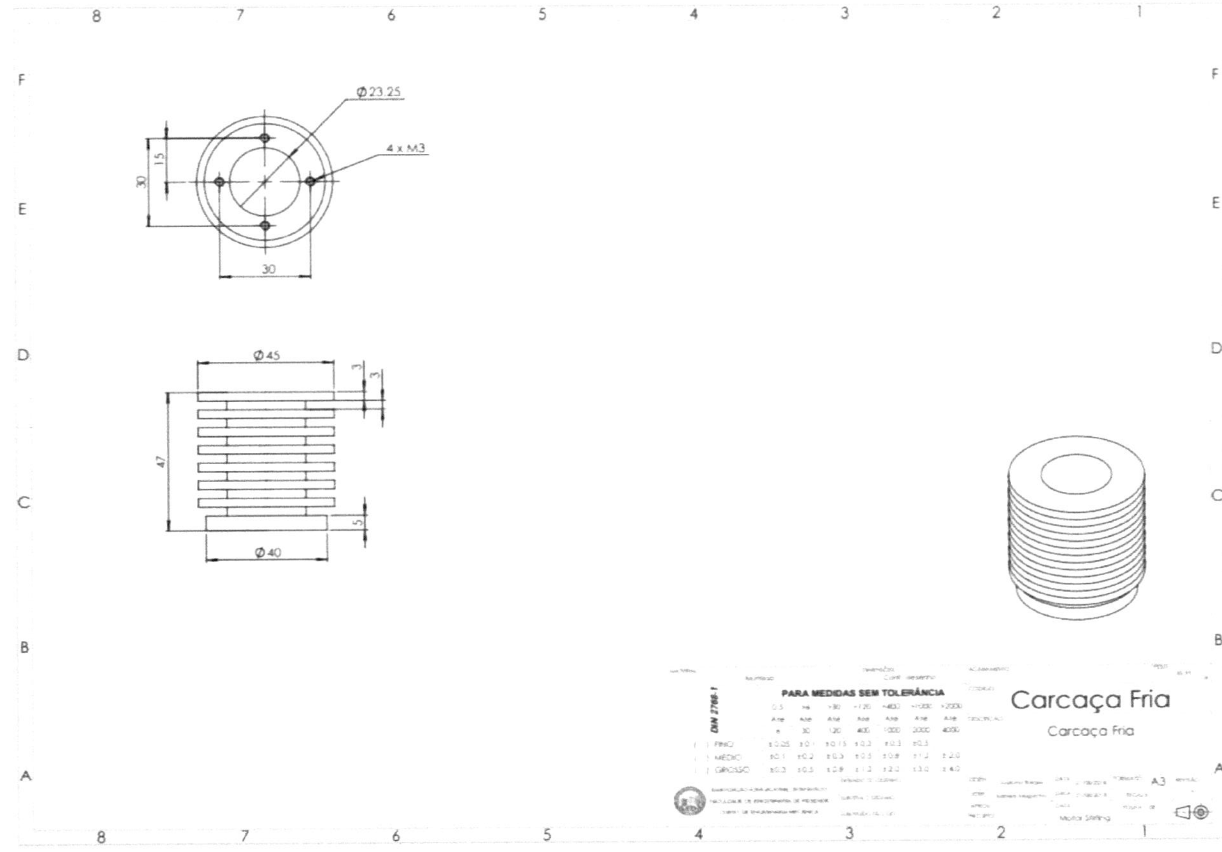

47
Ø40
Ø45
5
3
3
30
15
30
Ø23,25
4 x M3
PARA MEDIDAS SEM TOLERÂNCIA
DIN 2768-1
Carcaça Fria
Carcaça Fria
A3
51

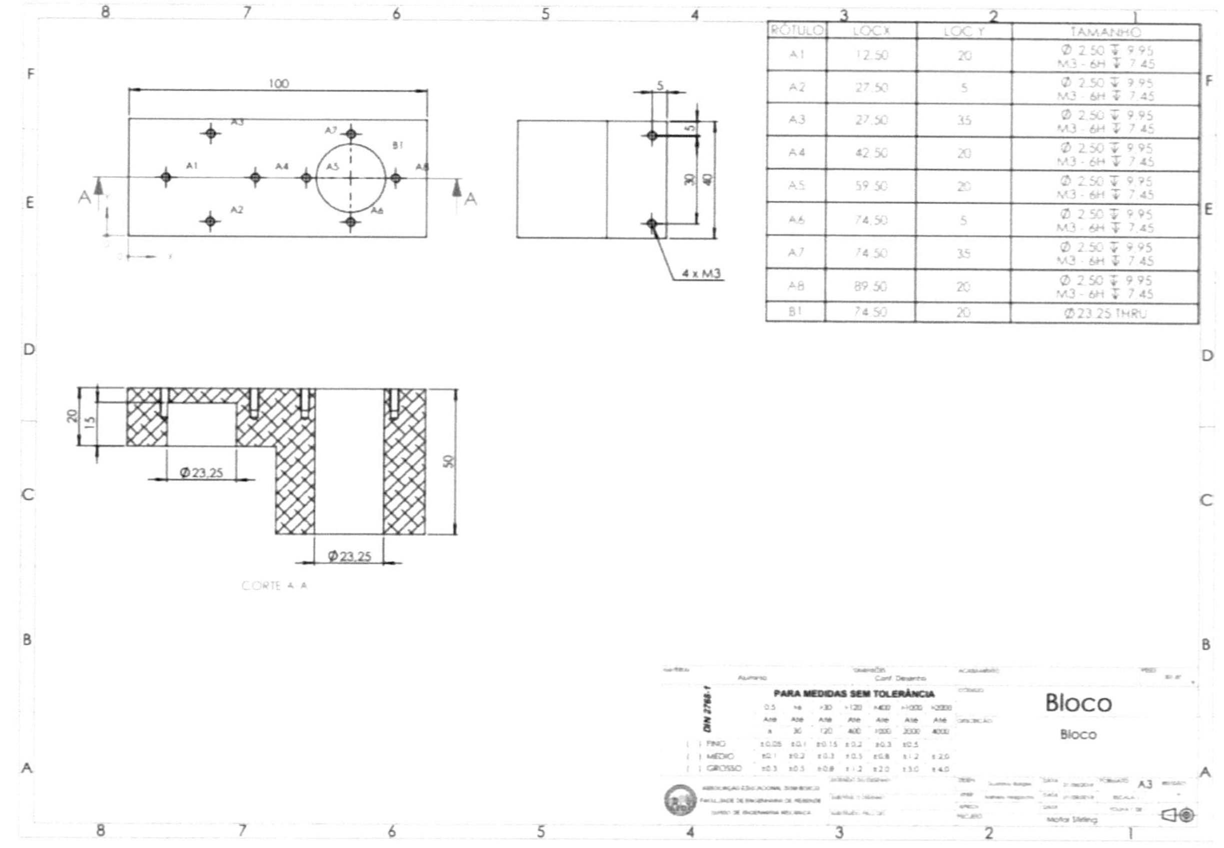

RÓTULO	LOC X	LOC Y	TAMANHO
A1	12.50	20	Ø 2.50 ▼ 9.95 M3 - 6H ▼ 7.45
A2	27.50	5	Ø 2.50 ▼ 9.95 M3 - 6H ▼ 7.45
A3	27.50	35	Ø 2.50 ▼ 9.95 M3 - 6H ▼ 7.45
A4	42.50	20	Ø 2.50 ▼ 9.95 M3 - 6H ▼ 7.45
A5	59.50	20	Ø 2.50 ▼ 9.95 M3 - 6H ▼ 7.45
A6	74.50	5	Ø 2.50 ▼ 9.95 M3 - 6H ▼ 7.45
A7	74.50	35	Ø 2.50 ▼ 9.95 M3 - 6H ▼ 7.45
A8	89.50	20	Ø 2.50 ▼ 9.95 M3 - 6H ▼ 7.45
B1	74.50	20	Ø 23.25 THRU

PARA MEDIDAS SEM TOLERÂNCIA

DIN 2768-1

	0.5	>6	>30	>120	>400	>1000	>2000
	Até	Até	Até	Até	Até	Até	Até
	6	30	120	400	1000	2000	4000
FINO	±0.05	±0.1	±0.15	±0.2	±0.3	±0.5	
MÉDIO	±0.1	±0.2	±0.3	±0.5	±0.8	±1.2	±2.0
GROSSO	±0.3	±0.5	±0.8	±1.2	±2.0	±3.0	±4.0

Bloco

Bloco

A3

Motor Stirling

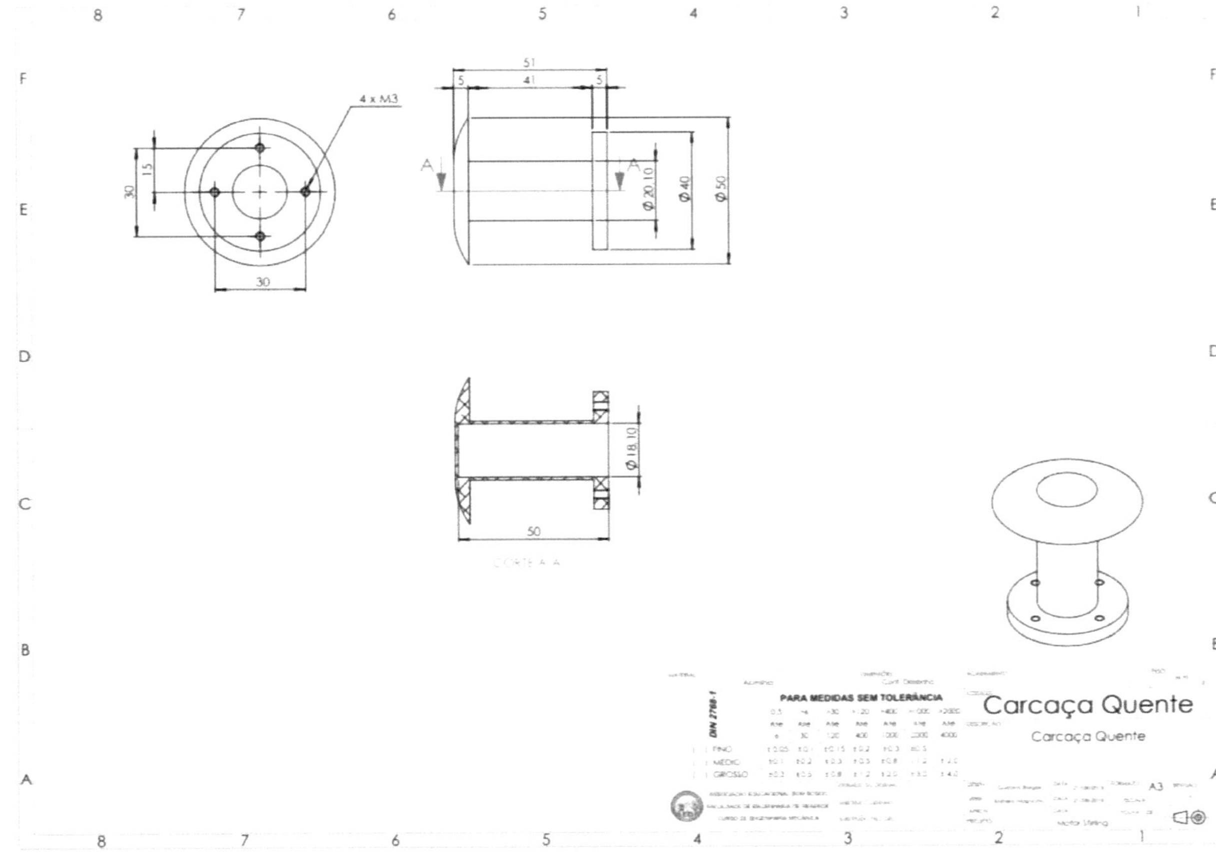

4 x M3
30
15
30
51
5
41
5
Ø 20,10
Ø 40
Ø 50
Ø 18,10
50
CORTE A-A
PARA MEDIDAS SEM TOLERÂNCIA
DIN 2768-1
Alumínio
FINO
MÉDIO
GROSSO
Carcaça Quente
Carcaça Quente
Motor Stirling
A3

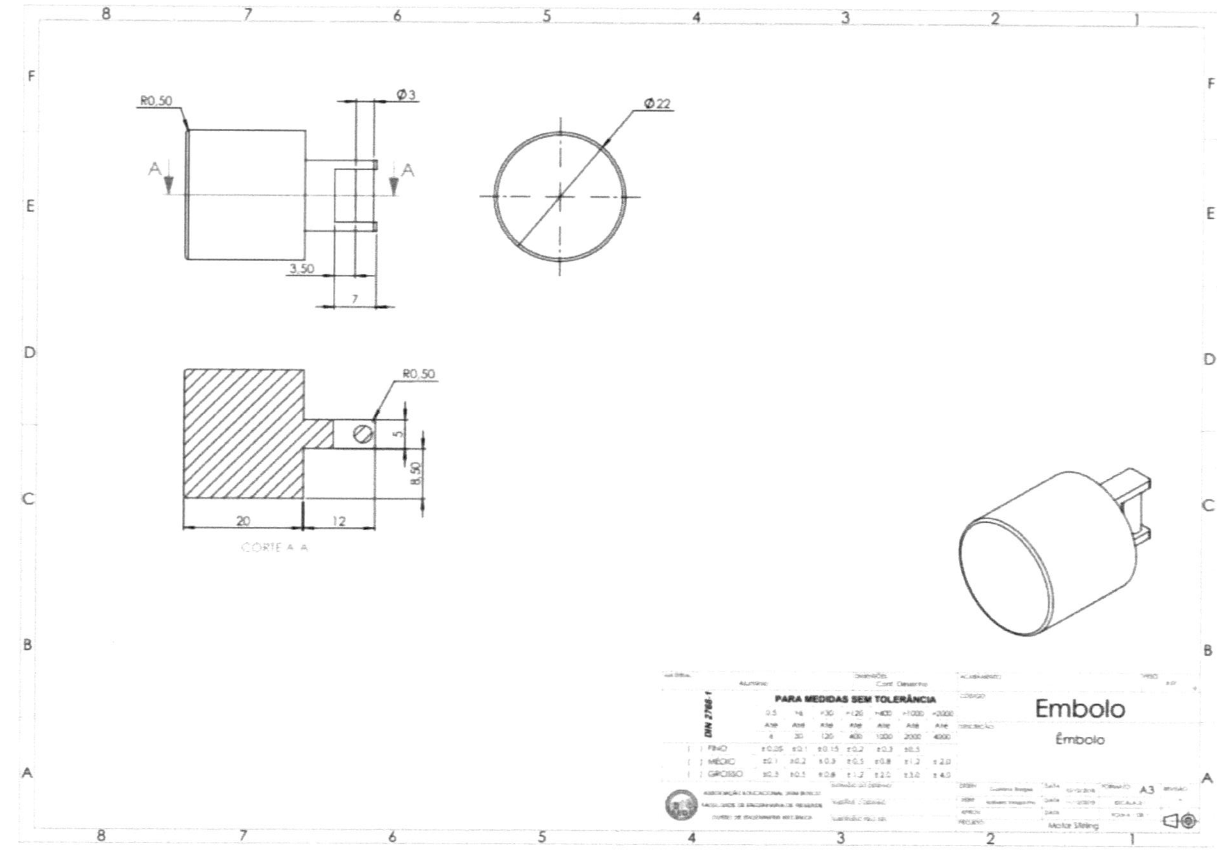

R0,50
Ø3
Ø22
A
A
3,50
7
CORTE A-A
R0,50
20
12
DIN 2768-1
PARA MEDIDAS SEM TOLERÂNCIA
Embolo
Embolo
A3

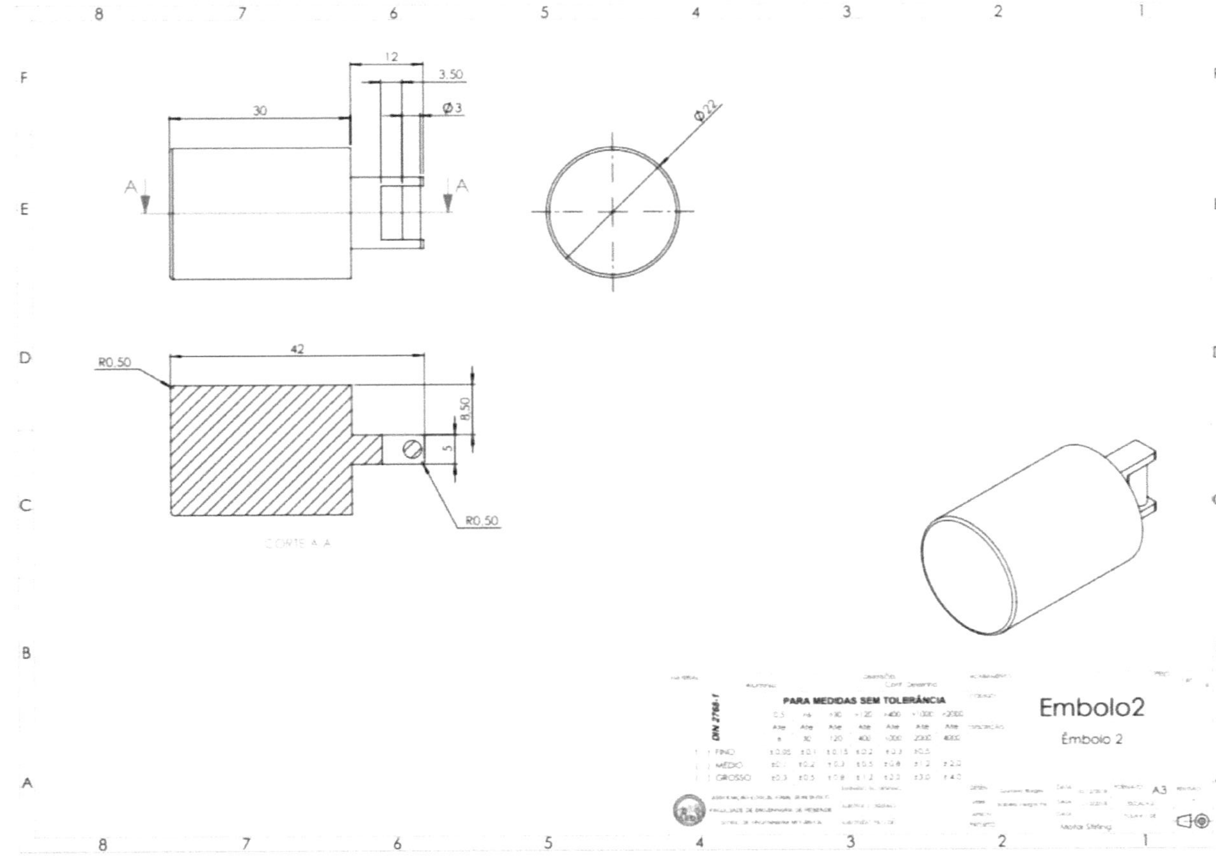

12
3.50
30
Ø3
Ø22
42
R0.50
8.50
5
R0.50
CORTE A-A
A
A
PARA MEDIDAS SEM TOLERÂNCIA
DIN 2768-1
Embolo2
Êmbolo 2
A3

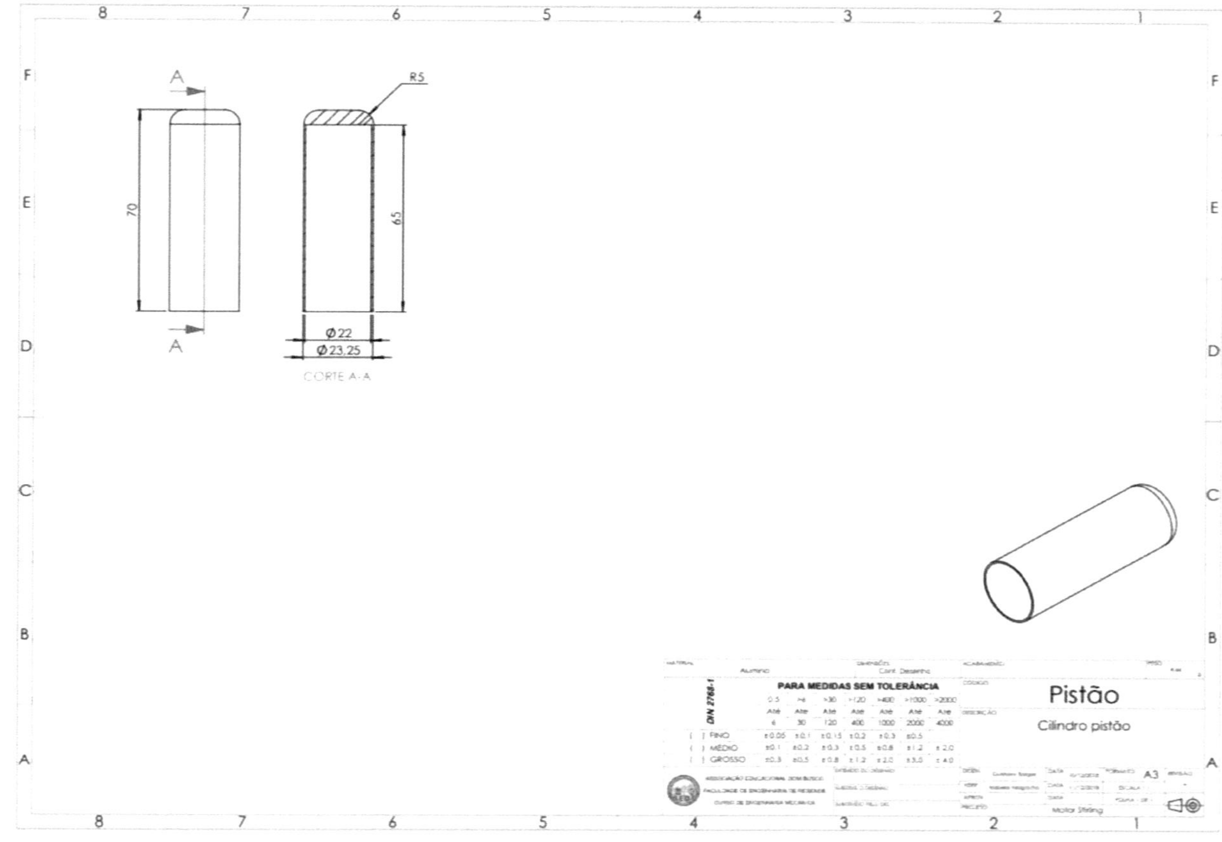

70
65
Ø22
Ø23,25
R5
CORTE A-A
A
A
PARA MEDIDAS SEM TOLERÂNCIA
DIN 2768-1
FINO
MÉDIO
GROSSO
Pistão
Cilindro pistão
A3
Motor Stirling
56

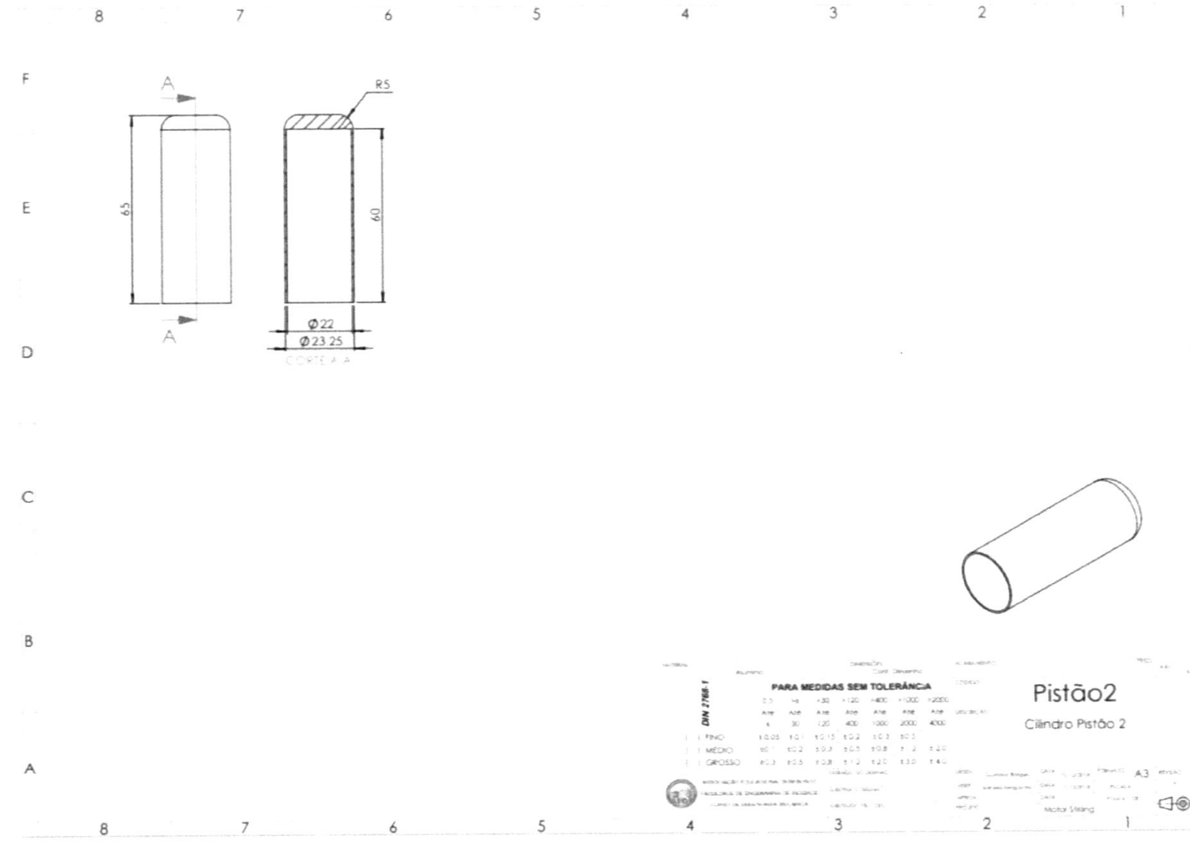

65
60
R5
Ø22
Ø23.25
CORTE A-A
A
A
PARA MEDIDAS SEM TOLERÂNCIA
DIN 2768-1
FINO
MÉDIO
GROSSO
Pistão2
Cilindro Pistão 2
A3

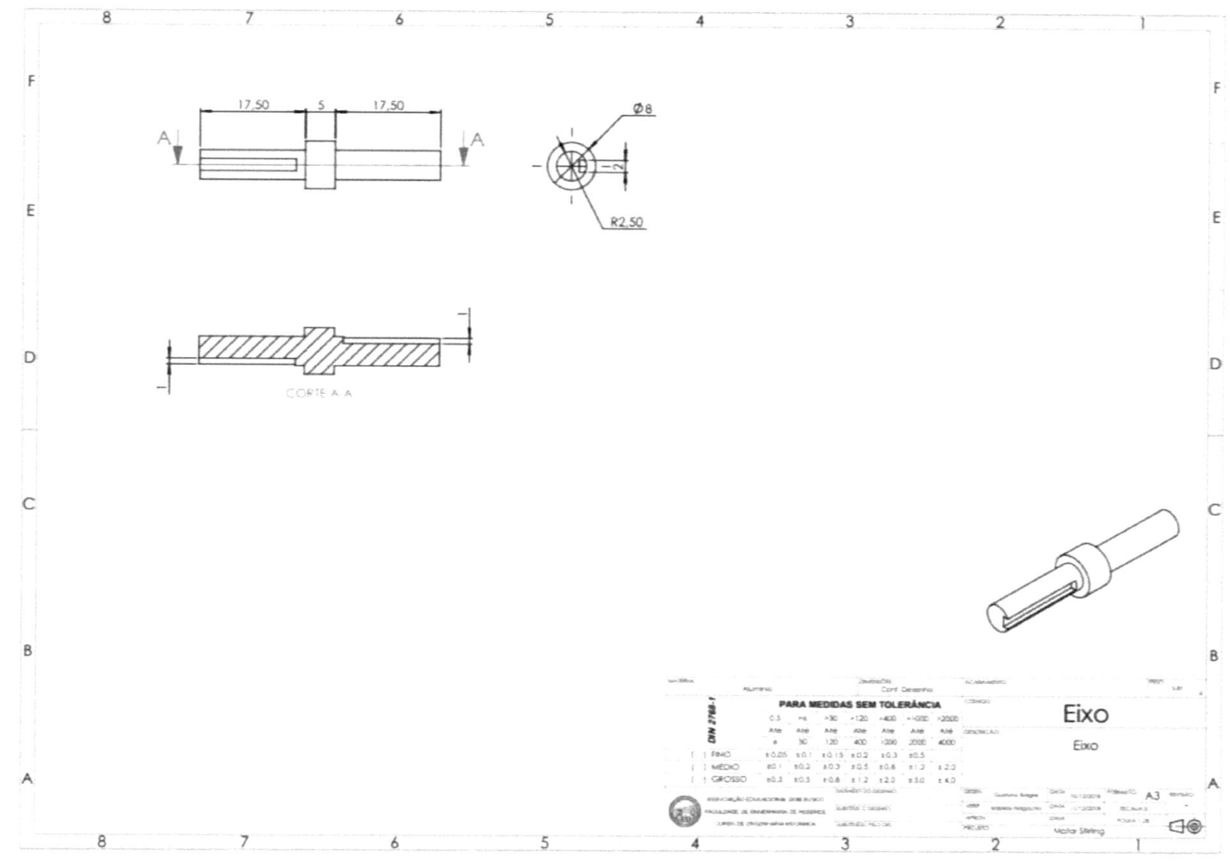

17,50   5   17,50
A   A
Ø8
R2,50
CORTE A-A
DIN 2768-1
PARA MEDIDAS SEM TOLERÂNCIA
0.5 >6 >30 >120 >400 >1000 >2000
Até Até Até Até Até Até Até
6 30 120 400 1000 2000 4000
FINO   ±0.05 ±0.1 ±0.15 ±0.2 ±0.3 ±0.5
MÉDIO   ±0.1   ±0.2 ±0.3   ±0.5 ±0.8 ±1.2 ±2.0
GROSSO ±0.3   ±0.5 ±0.8   ±1.2 ±2.0 ±3.0 ±4.0
ASSOCIAÇÃO EDUCACIONAL DOM BOSCO
FACULDADE DE ENGENHARIA DE RESENDE
CURSO DE ENGENHARIA MECÂNICA
Eixo
Eixo
A3
Motor Stirling

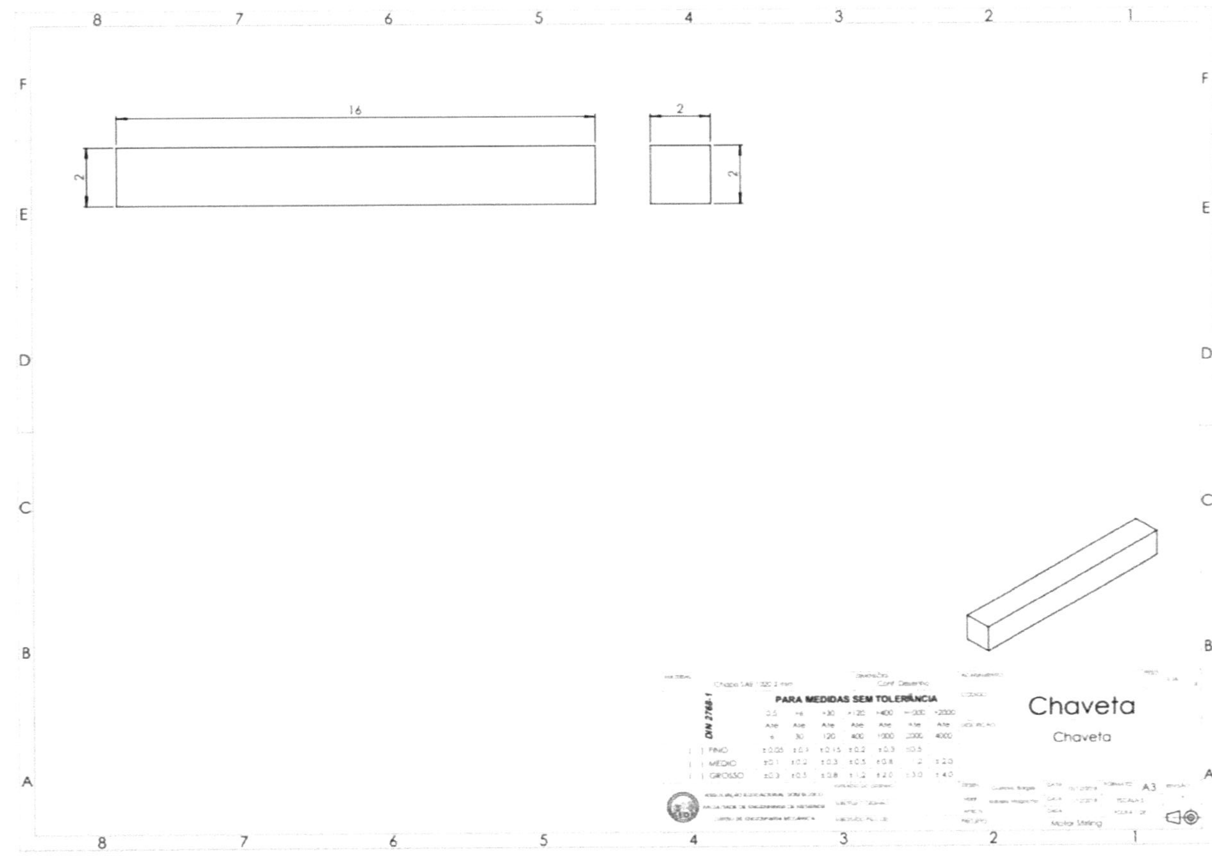
16
2
2
2
PARA MEDIDAS SEM TOLERÂNCIA
DIN 2768-1
FINO
MEDIO
GROSSO
Chaveta
Chaveta
A3
Conf. Desenho
59

Printed by Books on Demand GmbH, Norderstedt / Germany